Interactive Linear Algebra

Interactive Linear Algebra and its associated software are course materials for introductory linear algebra. The software consists of around 75 (and growing) highly visual online apps that are fully self-contained. Immersed in each app, students make problem-solving decisions in real time, while the software handles routine, low-level calculations to show the outcome of each decision. Students are unencumbered by tedious "pencil and paper" calculations that STEM professionals haven't used for decades, and focus on rapid problem-solving skills. Traditional exercises involving technology simply refer students to "black box" computational software, while the *Interactive Linear Algebra* software keeps students fully engaged in the mathematical processes they are learning. This engagement leads to better retention of problem-solving strategies and deeper mathematical insight.

Interactive Linear Algebra includes standard introductory topics from basic matrix algebra to Singular Value Decomposition. The development of these topics is mathematically rigorous, but the proofs are often inspired by mathematical insights that are imbued by the software. This strategy allows instructors to choose an emphasis on formal proofs that matches the learning objectives of their course.

- Covers all standard topics in basic linear algebra in a mathematically rigorous way.
- Permits a broad range of emphasis on mathematical logic and proofs.
- Allows students to carry out problem-solving steps in real time via immersive apps that handle routine arithmetic and immediately provide the numerical outcome of each student's decisions.
- Promotes retention of problem-solving strategies and mathematical insight.
- Avoids the need for computational software training in class.
- Saves significant class time, replacing tedious hand-calculations by instructors with brief demonstrations of the self-contained apps.

Interactive Linear Algebra is intended for use in introductory linear algebra courses for students pursuing STEM degrees. The text allows any level of emphasis on mathematical rigor, from light treatment to a formal introduction to mathematical proofs.

Interactive Linear Algebra

Conrad Plaut

CRC Press
Taylor & Francis Group
Boca Raton   London   New York

CRC Press is an imprint of the
Taylor & Francis Group, an **informa** business

A CHAPMAN & HALL BOOK

First edition published 2026
by CRC Press
2385 NW Executive Center Drive, Suite 320, Boca Raton FL 33431

and by CRC Press
4 Park Square, Milton Park, Abingdon, Oxon, OX14 4RN

CRC Press is an imprint of Taylor & Francis Group, LLC

ISBN: 978-1-041-14424-3 (hbk)
ISBN: 978-1-041-14417-5 (pbk)
ISBN: 978-1-003-67436-8 (ebk)

DOI: 10.1201/9781003674368

Typeset in Latin Modern font
by KnowledgeWorks Global Ltd.

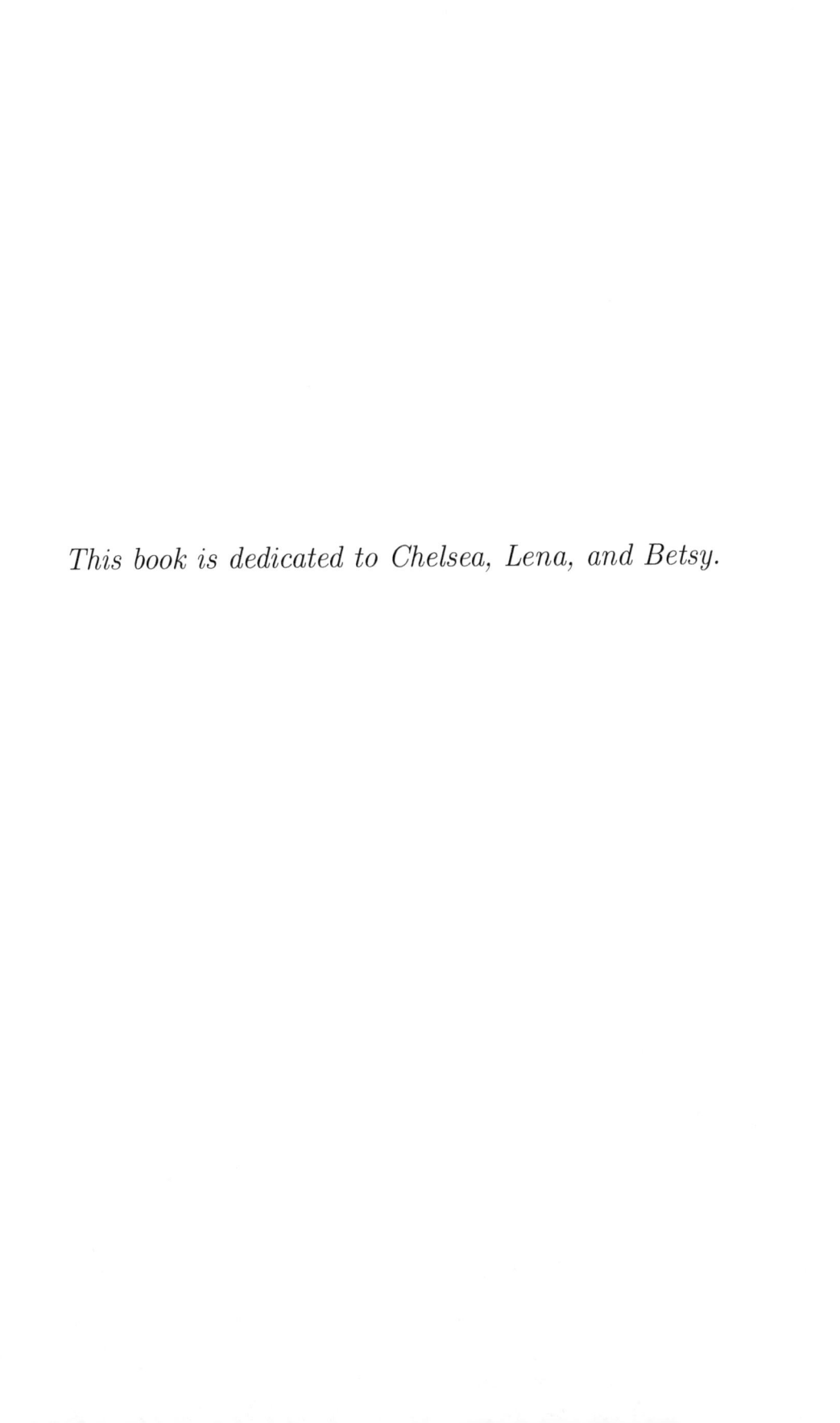

This book is dedicated to Chelsea, Lena, and Betsy.

Contents

Preface   ix

Author   xv

1 Matrices and Matrix Algebra   1
   1.1 Basic Matrix Definitions and Operations   1
   1.2 Gauss-Jordan Elimination   3
   1.3 Algebraic Operations on Matrices   7
   1.4 Algebraic Properties of Matrix Operations 11
   1.5 The Trace and the Transpose 17
   1.6 Inverses of Matrices 20

2 Systems of Linear Equations and Matrices   24
   2.1 Matrix Form of Systems 24
   2.2 Solving Systems Using Gauss-Jordan Elimination 28
   2.3 Applications of Linear Systems 34
   2.4 Matrix Form for Systems of Linear Equations 39
   2.5 Homogeneous Systems 41
   2.6 Parameterized Solutions 46
   2.7 Elementary Matrices 50
   2.8 Applications of Elementary Matrices 54

3 The Determinant   58
   3.1 Basic Properties and Elementary Matrices 58
   3.2 Cofactor Expansion 69

4 Euclidean Vector Spaces   75
   4.1 Vectors in Dimension 2 75
   4.2 Vectors and Norms in Euclidean Space 82
   4.3 Lines Determined by Points and Vectors 89
   4.4 Orthogonality 93
   4.5 The Cross Product 97
   4.6 Matrices as Linear Transformations 100
   4.7 Rotations in the Plane 105
   4.8 Composition of Linear Transformations 107
   4.9 Geometric Meaning of the Determinant 108

5 Vector Spaces **113**
5.1 Vector Spaces and Subspaces 113
5.2 Linear Transformations 119
5.3 Linear Combinations and Linear Independence 121
5.4 Spanning Sets 125
5.5 Bases 126

6 Finite Dimensional Vector Spaces **132**
6.1 Dimension 132
6.2 Multilinearity of the Determinant 138
6.3 Geometric Interpretation of Dimension and Subspaces 144
6.4 Change of Coordinates 144
6.5 Matrix Expression of a Linear Transformation 146
6.6 Row and Column Spaces of a Matrix 150
6.7 Null Space of a Matrix 153
6.8 Orthonormal Bases and Orthogonal Matrices 155
6.9 Orthogonal Subspaces and Gram-Schmidt 159
6.10 The Structure of the Row, Column and Null Spaces 163

7 Eigenvalues and Eigenvectors **165**
7.1 Eigenvalues and Eigenvectors 165
7.2 The Structure of the Characteristic Polynomial 168
7.3 Diagonalization 173
7.4 Equivalent Conditions to be Invertible 178
7.5 Complex Numbers with Applications to Eigenvalues 179
7.6 The Spectral Theorem and SVD 185

Index **191**

Preface

Many elementary linear algebra texts used in contemporary sophomore-level courses are only superficially different from those used in the mid-20th century, when the word "computer" referred to a person. The texts of that time were written to prepare future STEM professionals for the daily reality of rapid, accurate by-hand calculations. These days, STEM professionals almost never do such calculations by hand (which would be impossible for most modern problems anyway), yet the 1970s pedagogical strategy lives on. For example, the text that I used in the fall of 2022, selected by my department, was first published in 1973. It bears a striking resemblance to the text (by a different author) that was used when I took the course as an undergraduate in 1981. That "updated" 1973 book that I used in 2022 has modern touches like pictures of "eigenfaces" and supplemental materials designed to be used with external computational software. Yet the text, including most of the exercises and examples, is vintage mid-1900s. Students are required to practice tedious by-hand calculations for each new concept. When asking for a solution "by inspection", the text helpfully explains that this means "without pencil and paper". OK, so can you use your iPad? Cramer's Rule, which is woefully inefficient for large matrices, is still a core topic. Determinants are introduced via cofactor expansion—also a spectacularly inefficient method for the large matrices that are common in modern science.

Deferring problems to external computational software also has issues and involves pedagogical risks. For one thing, this is a bit of work for the student—they have to learn to use the particular software that is being used for the class, enter data (without error), run the software, and then copy back the answer (without error). Instructors also must use valuable time to ensure that students can use the software and deal with the inevitable questions and technological issues. But the most important risk is that students will not learn the underlying processes for the tools they are learning to use. The black box has done its work, but has the student learned anything meaningful? Does the student now understand the properties of the determinant, how it depends on properties of the matrix, how it may be computed, and so on? *Interactive Linear Algebra* and its associated software preserve one of the most important pedagogical requirements to understand mathematics and its applications, namely, to understand the *processes* through which the mathematics does its work. The software has around 75 (and growing) highly visual interactive online apps called *Math Pauses* (MPs) that allow students to work through basic processes, while the software handles what is often simply tedious 5th-grade arithmetic. For example, one of the core MPs allows students to perform the row operations of Gauss-Jordan elimination in a highly visual manner, including dragging and dropping scalar multiples of rows. The app does basic arithmetic and displays the result of the row operation suggested by the student. The student can then immediately look at the result, make the next decision, or "undo" the row operation if they don't like what they see. This core process is then used again and again for various purposes throughout the course, including as the starting point for determinants. Later in the course, when dealing with more complex problems, students may be permitted to simply invoke Gauss-Jordan elimination as an "earned" tool so that they can focus on new, more complex processes that incorporate it.

I have heard the question: isn't something of pedagogical value lost if we do not *allow* students to make mistakes so they may learn from them? But what does a student "learn" from making a 5th-grade arithmetic error? Students in a sophomore-level math class do know how to do 5th-grade arithmetic by hand, albeit perhaps not as well as a college student in the 1970s. They can do it, but they occasionally make a mistake. Who doesn't? After making a trivial mistake, the student must at some point take time to find the error and redo the work, creating frustration and slowing down the problem-solving process. In fact, the problem-solving process is slowed even if the student *doesn't* make a mistake. Isn't it better for students to make rapid problem-solving decisions (the important part!) uninterrupted by the sort of trivial calculations that STEM professionals haven't done by hand for decades?

For example, adding a scalar multiple of a row to another in a 3×3 matrix requires three multiplications, three sums, and rewriting the matrix. What is "learned" from doing that arithmetic, which might need to be done flawlessly a dozen times to reduce the matrix to reduced echelon form? In contrast, using the Gauss-Jordan Elimination MP, students typically learn to entirely reduce a 3×3 matrix to reduced echelon form in less than a minute or two by developing their own efficient strategies. As a result, students can do large numbers of exercises rapidly, focusing on problem-solving decisions rather than on whether or not they got that minus sign right. Those who think students should still be doing *some* 5th-grade math may be comforted by the fact that the software only permits fractional entry, requiring some mental fractional arithmetic. After all, grabbing a calculator disrupts the flow as much as grabbing a pencil, and fractions have more intuitive content than decimals. But if students do make an error in their mental fractional arithmetic, they will find out immediately, so they can try again, rather than continuing blithely onward to a wrong answer.

To be clear: The software is not simply a computational crutch for students. Allowing students to make rapid problem-solving decisions and immediately see the calculated output of those decisions can help build mathematical instincts—and can even illuminate the connection between mathematical processes and mathematical proofs.

Speaking of proofs, at this point some instructors reading this Preface may be concerned (and some students may be hoping) that *Interactive Linear Algebra* avoids proofs. Quite the opposite is true. *Interactive Linear Algebra* pays more careful attention to logical development and proofs than many sophomore-level texts, which sometimes indulge in hand-waving and even have logical errors. Basic linear algebra is sometimes used as a vehicle to introduce students to logic and proof, because proofs in the subject naturally progresses from short and simple to relatively complex. *Interactive Linear Algebra* takes advantage of this progression to gradually increase level of complexity of proofs, as part of a complete logical development of the subject, culminating with proofs of the Spectral Theorem for real symmetric matrices and Singular Value Decomposition. But instructors using the text are free to expect from their students a level of proof understanding that is appropriate for the learning objectives of their course and the preparation of their students. And for a student with an interest in mathematical proofs who is in a class that treats them relatively lightly, the proofs are there for them to work through, perhaps for honors credit. To help with learning proof techniques, there are many "finish the proof" exercises that require students to fill in some relatively straightfoward details that were left out of a proof. These exercises encourage students to read the proof to orient themselves before starting the exercise.

Now a word for students who might be disappointed or apprehensive about the prospect of learning about proofs. Most proofs in elementary linear algebra courses are very "algorithmic" in the sense that they are often easily translatable into methods to compute what the theorem is about. Modern students majoring in sciences and engineering are almost always required to learn some basic programming because modern scientific and engineering applications often require dealing with code–modifying it for specific applications if not writing it from scratch. Often students are asked to code the processes that they are learning, and there is good pedagogical reason to do so: coding the process helps with understanding. Likewise, learning basic proofs helps with understanding of mathematical ideas. In fact, computer code is basically a "proof that does something", using a very specialized language with its own syntax.

Here are a few practical considerations for instructors who are adopting *Interactive Linear Algebra* and the associated software for a linear algebra course. Updated information on the software, including current links to the MPs for those using a paper copy of this book is available at plautpedagogy.com. The number of MPs will continue to expand as I find new ways to adapt additional exercises to the MP format. It is highly recommended that instructors assign *all* of the MPs as the class progresses through *Interactive Linear Algebra,* with due dates very soon after the corresponding material in *Interactive Linear Algebra* is covered. In fact, MPs should be assigned nearly every day, so that students learn the processes quickly and are prepared for what comes next. The MPs are listed in the text at the most appropriate points for them to be assigned, with numbering and names matching how they are listed in the software to make them easy to find. Students will master the MPs faster than one might expect, and assigning 4-6 per week is quite reasonable. Each MP covers a specific mathematical process, and there are several "repetitions" for each MP.

MPs are automatically graded by *completion*, which is possible because problem-solving steps must be entered by the student in real time. This feature should be contrasted with the automatic grading of typical Learning Management Systems, which are only able assess *answers*. Assessment of (only) answers can be a source of frustration for students, who are told their answer is wrong, but not exactly why, and must hunt through their solution for what is likely a 5th-grade arithmetic error. Software systems that only check *answers* also greatly increase the temptation to cheat. Cheating via external sources is bad for everyone: those who do it (who don't learn) and those who don't (who may be unfairly disadvantaged in terms of their grades). Generative AI has made cheating in standard mathematics courses far simpler and more efficient. Even complete solutions are easy to obtain. For exercises in a class at this level, the AI solutions are more or less accurate. This puts tremendous pressure on students who would still like to do their work honestly, but fear they will lose out in terms of their grades. Yet cheating on an MP would be so inefficient as to hardly be worthwhile as a time-saver. Students must make problem-solving decisions in real time and submit the parameters of each decision. Utilizing even an entire solution found online would be cumbersome, error-prone (to enter in real time), and time-consuming. Online solutions likely involve decimal expressions, which may not be entered in MPs. Students will find it easier to just take a little time to learn to solve the problems quickly! In fact, students seem to greatly enjoy using the MPs, which remove a significant amount of tedium and frustration from their work.

There are also many written exercises in the text, which may be assigned and graded by the instructor according to their courseload and course objectives.

As for exams, by summer 2026 the software will have functionality to allow instructors to create exams using MPs of their choice, which will have an "exam mode" that is modified from the original by removing hints or instructions. When I last taught with the materials, I gave hybrid exams with an MP portion lasting up to 25/50 minutes, followed by a written portion that I graded by hand. The MP portion of the test is self-grading based on completion, and the instructor will receive a report for each student listing their completion success or partial success. The latter may be used to offer partial credit if the instructor so chooses. Again, unlike multiple-choice exams, MP-based exams grade students based on their *decisions*, not just their answers.

The standard topics are all in *Interactive Linear Algebra* (albeit with only a glance in the direction of Cramer's Rule), but there are changes in the way the topics unfold. Chiefly among them is that much of the course is built around the core Gauss-Jordan elimination process. Perhaps the biggest resultant change is that determinants are not introduced in the traditional way, via cofactor expansion. Instead, the determinant is introduced as a function satisfying properties that correspond to the row operations in Gauss-Jordan elimination. Students may then immediately begin computing determinants using a process they already know, entering the factors that change the determinant with each row operation. Among other things, this makes it easy for students to visually understand the advantages of only reducing to upper triangular form, and also motivates the proof that the determinant is the product of the diagonal elements of an upper triangular matrix. Gauss-Jordan elimination also helps students understand why the determinant is the unique function with those properties, which is proved rigorously in the text.

Using cofactor expansion as the definition of the determinant may seem more comfortable for students because for small matrices there is a "formula". But the reality is that the definition via cofactor expansion involves a process, not a formula. Focusing on processes and not formulas furthers what has been a goal of freshman calculus and other lower-division math courses for at least three decades, namely to break down the problematic misconception of many students that functions *are* formulas. In *Interactive Linear Algebra*, students already know the determinant, its properties and its uniqueness long before learning, via uniqueness and multilinearity, that cofactor expansion is an alternative (albeit incredibly inefficient) way to compute it. Cofactor expansion definitely has a place in elementary linear algebra. In *Interactive Linear Algebra*, beyond providing an introduction to iterative definitions, there are two main *raisons d'être* for cofactor expansion. First, it is used to show that the absolute value of the determinant of a matrix is the volume of the parallelepiped spanned by its columns (something that cannot easily be seen via Gauss-Jordan elimination). Second, cofactor expansion is essential to develop fundamental properties of the characteristic polynomial.

Multilinearity, which is often given the undignified name "row additivity" in elementary texts and treated lightly if at all, is a fundamental concept that recurs in many areas of science, engineering and mathematics. *Interactive Linear Algebra* strives to do justice to this important topic.

Finally, *Interactive Linear Algebra's* MPs to learn about parameterized solutions, LU-decomposition, eigenvalues and eigenvectors take full advantage of the processes that students have learned to allow them to do far more (and more complete) exercises than is feasible "by hand", while allowing students to go through the main steps of the entire process, utilizing many of the tools that they have learned throughout the semester–all without having to access external "black box" software that hides the processes that they must learn to understand.

Author

Conrad Plaut is a mathematics professor and former department head at the University of Tennessee. He earned his BA from Guilford College in Mathematics and English, after which he accepted a Fulbright Scholarship to study in Zagreb, Croatia. Plaut earned his PhD in Mathematics from the University of Maryland. At the University of Tennessee he has taught more than two dozen different courses, from freshman calculus to graduate topics courses. Plaut has been awarded over $1.7 million in grants and awards, primarily from the National Science Foundation, to support his mathematical research, conferences he has co-organized, and programs he has organized for students. The latter include a scholarship program for students from the Appalachian region majoring in math and computer science, a Research Experiences for Undergraduates (REU) program, and the University of Tennessee Math Honors Program, for which he served as founding director. Plaut has given around 70 research talks in 11 countries and has over 700 citations to his research papers (source: Google Scholar).

1

Matrices and Matrix Algebra

In this chapter you will learn about matrices and basic operations on matrices. The goal is to become proficient at using the MPs apps as preparation to learn *why* you have learned these operations. Linear Algebra has applications in all scientific fields, and being proficient in the main mathematical processes is very important for anyone planning a career in science, engineering, or mathematics.

1.1 Basic Matrix Definitions and Operations

A matrix is a rectangular array of numbers. We will sometimes, especially when doing row operations, refer to numbers by their more formal name, "scalars". In fact, nearly all of what you will be learning about matrices works when the scalars are something other than the real numbers—most importantly complex numbers. But there are also times when one wants to be more restrictive—for example, using only integers for scalars. Until much later on, we will only use the real numbers $\mathbb{R}$ as scalars.

If a matrix $\mathbf{A}$ has m rows and n columns, $\mathbf{A}$ is called an $m \times n$ matrix, or a matrix of *size* $m \times n$. Here is an example of a 3×4 matrix with real number scalars:

$$\begin{bmatrix} 1 & -1 & \pi & 2 \\ 0 & 2/3 & 1 & 12 \\ 4 & 5 & 1/6 & 7 \end{bmatrix}$$

The ordering of these numbers is essential: you can remember "rows-by-columns". A matrix $\mathbf{A}$ is called "square" if it has the same number of rows and columns—i.e. A is an $n \times n$ matrix. Below is the first Math Pause, abbreviated as MP. The intention is for you to not proceed with these materials until you have worked through all of the individual repetitions in each MP. If you have not mastered each MP, it may be more difficult to learn what comes after. After you have worked through any MP, you may always return for more practice as needed.

DOI: 10.1201/9781003674368-1

MP 1 *MATRIX SIZE*

The individual numbers of the matrix are called *entries* or *components*. Each entry of a matrix **A** may be described by the row and column in which it "lives". That is, the entry in the ith row and j^{th} column of **A** is denoted a_{ij} or $\mathbf{A}_{ij}$. Why two different ways of writing the same thing? Linear algebra involves complex calculations, which are often easier to represent using one notation or another—and different people use different commonly used notations! So it is important to remain "notation flexible" and be fluent in the language of linear algebra, regardless of the specific notation used. Typically matrices are denoted in this way: by bold capital letters with their entries denoted by the corresponding letters not in bold. For example, $\mathbf{A} = [a_{ij}]$.

MP 2 *MATRIX ENTRIES*

MP 3 *MATRIX ENTRIES II*

There are some special terms for square matrices. The *diagonal* of a square matrix consists of the entries a_{ii}, which are simultaneously in the ith row and the ith column. A square matrix is *upper triangular* if all of the entries below the diagonal entries are 0. For example

$$\begin{bmatrix} 2 & -1 & 4 \\ 0 & 9 & 1 \\ 0 & 0 & 5 \end{bmatrix}$$

is upper triangular. Put another way, a square matrix $[a_{ij}]$ is upper triangular if $a_{ij} = 0$ whenever $j < i$. A square matrix is *lower triangular* if all entries above the diagonal are 0 (equivalently, $a_{ij} = 0$ whenever $i < j$). Note that the diagonal entries in any upper or lower triangular matrix can also be 0. For example

$$\begin{bmatrix} 0 & 0 & 0 \\ 1 & 3 & 0 \\ 1 & 0 & 5 \end{bmatrix}$$

is lower triangular. A square matrix that is *both* upper and lower triangular has entries equal to zero above and below the diagonal. Put another way, the only possible non-zero entries are on the diagonal:

$$\begin{bmatrix} -2 & 0 & 0 \\ 0 & 1 & 0 \\ 0 & 0 & 5 \end{bmatrix}$$

This kind of matrix is called a *diagonal* matrix.

Another very important type of matrix is *symmetric matrices*, which have the property that when you reverse their rows and columns, the matrix doesn't change. That is, they are symmetric about their diagonals. Here is an example of a symmetric matrix.

$$\begin{bmatrix} 1 & 2 & 3 & 4 \\ 2 & 5 & 6 & 7 \\ 3 & 6 & 8 & 9 \\ 4 & 7 & 9 & 0 \end{bmatrix}$$

MP 4 *MATRIX PROPERTIES*

1.2 Gauss-Jordan Elimination

The first matrix process you will learn is Gauss-Jordan elimination. This process involves three *row operations* to reduce a matrix to a more useful form. One of the most useful forms is known as *reduced echelon form.* To describe this form, first define the *leading entry* of a row to be the first (from left to right) entry that is not zero. If a row has all 0s (which we will refer to as a 0-*row*), then it has no leading entry.

MP 5 *LEADING ENTRIES*

A matrix is in *reduced echelon form* if

1. All rows without a leading entry (i.e. all entries are zero) are at the bottom of the matrix.

2. All leading entries are 1.

3. The leading entry of any row is to the right of the leading entries of rows above it (this appears like an "upside-down staircase").

4. All other entries in the column of a leading entry are 0s.

For example, the following matrix is in reduced echelon form:

$$\begin{bmatrix} 1 & 2 & 0 & 0 & 2 \\ 0 & 0 & 1 & 0 & 3 \\ 0 & 0 & 0 & 1 & -1 \\ 0 & 0 & 0 & 0 & 0 \\ 0 & 0 & 0 & 0 & 0 \end{bmatrix}$$

As you should check for yourself, only the top three rows have leading entries, and the leading entry of any row is to the right of the leading entries above it. In each column with a leading entry, there are only 0s above and below it.

MP 6 *MATRIX PROPERTIES II*

One important goal of Gauss-Jordan elimination is to transform a matrix into reduced echelon form, although in some circumstances it can be useful to stop at an earlier point (for example, reducing a square matrix to upper triangular form—something you will learn about later). You will learn later several reasons why Gauss-Jordan elimination is useful.

Gauss-Jordan elimination involves repeatedly using three row operations on a given matrix:

RO1 **Multiplying a row by a non-zero scalar.**

RO2 **Adding a scalar multiple of one row to another.**

RO3 **Swapping two rows.**

In a standard linear algebra class, students are still expected to carry out these operations by hand or to enter matrices into a separate computational engine. Doing Gauss-Jordan elimination by hand does help students learn to understand this important process, but the task involves a large number of tedious calculations. This tedium contributes very little to the student's understanding and can lead to frustrating arithmetic errors (including under the pressure of taking a test). I will illustrate this process by *partially* reducing the following matrix to reduced echelon form. You may only wish to skim through this discussion, finding it easier to start with the Gauss-Jordan Elimination MP as soon as possible.

$$\begin{bmatrix} 5 & 2 & 0 \\ 2 & -2 & 4 \\ 3 & 1 & 2 \end{bmatrix}$$

For the first step we will use **RO2**, the most frequently used of the row operations: Add $-3/5$ times the top row to the bottom row. The goal is to change the leading entry of the third row to 0 as a first step in changing the matrix to reduced echelon form. First multiply the top row by $-3/5$, so it becomes $\mathbf{R} = \begin{bmatrix} -3 & -\frac{6}{5} & 0 \end{bmatrix}$. We do *not actually change the top row*, but rather we take the new row $\mathbf{R}$ and add its entries to the entries of the bottom row to get the new row $\begin{bmatrix} -3+3 & -\frac{6}{5}+1 & 0+2 \end{bmatrix} = \begin{bmatrix} 0 & \frac{1}{5} & 2 \end{bmatrix}$. We now replace the bottom row in the original matrix by this new row, so we now have the new matrix:

$$\begin{bmatrix} 5 & 2 & 0 \\ 2 & -2 & 4 \\ 0 & \frac{1}{5} & 2 \end{bmatrix}$$

As a next step we can add $\frac{-2}{5}$ times the top row to the second row, with the goal being to get a 0 in the first entry of the second row. Then we get the matrix:

$$\begin{bmatrix} 5 & 2 & 0 \\ 0 & -\frac{14}{5} & 4 \\ 0 & \frac{1}{5} & 2 \end{bmatrix}$$

You are encouraged to check this by hand (once!) so you see exactly what is happening, and also so that you can appreciate that the MP will do the arithmetic for you! Next, to zero out the second entry in the third row, we add $\frac{1}{14}$ times the second row to the bottom row to get the matrix

$$\begin{bmatrix} 5 & 2 & 0 \\ 0 & -\frac{14}{5} & 4 \\ 0 & 0 & \frac{16}{7} \end{bmatrix}$$

Finally we "finish" the bottom row by multiplying it by $\frac{7}{16}$ to get the matrix

$$\begin{bmatrix} 5 & 2 & 0 \\ 0 & -\frac{14}{5} & 4 \\ 0 & 0 & 1 \end{bmatrix}$$

We can now work upwards, adding -4 times the bottom row to the second row to "get rid of" the 4, and so on. In MP 7, the goal will be to reduce the 3×3 matrix to a very special matrix, called the "identity matrix". We will explain this terminology later, but the 3×3 identity matrix is

$$I_3 = \begin{bmatrix} 1 & 0 & 0 \\ 0 & 1 & 0 \\ 0 & 0 & 1 \end{bmatrix}$$

One final hint: as you surely have noticed, Gauss-Jordan elimination very quickly can involve arithmetic with fractions. Along these lines, MPs based on the Gauss-Jordan elimination app *do not permit decimal entries!* All scalars *must be fractions or whole numbers*. So rather than using your calculator, you will have to do some brief mental fractional arithmetic to decide how to proceed—but if you calculate wrong (and that can be easy to do at first), just hit the "undo" button and correct the mistake. Note that the MP also does not permit you to enter a "0" for the scalar. This is because (a) you may only multiply a row by a non-zero scalar and (b) adding 0 times one row to another row doesn't actually change the matrix!

MP 7 *GAUSS-JORDAN ELIMINATION*

The matrices that you just worked with were all chosen to reduce to the identity matrix as their reduced echelon form. Not all matrices have this property—in fact, Jordan-Gauss elimination can be applied to matrices that are not square, as we will discuss later. For now, let's examine the other possible reduced echelon forms for a 3×3 matrix. You have already seen the identity matrix, which has three leading entries. Below are the matrices with two leading entries, followed by those with one leading entry, where a and b

are scalars (possibly equal to 0):

$$\begin{bmatrix} 1 & 0 & a \\ 0 & 1 & b \\ 0 & 0 & 0 \end{bmatrix} \quad \begin{bmatrix} 1 & a & 0 \\ 0 & 0 & 1 \\ 0 & 0 & 0 \end{bmatrix} \quad \begin{bmatrix} 0 & 1 & 0 \\ 0 & 0 & 1 \\ 0 & 0 & 0 \end{bmatrix}$$

$$\begin{bmatrix} 1 & a & b \\ 0 & 0 & 0 \\ 0 & 0 & 0 \end{bmatrix} \quad \begin{bmatrix} 0 & 1 & a \\ 0 & 0 & 0 \\ 0 & 0 & 0 \end{bmatrix} \quad \begin{bmatrix} 0 & 0 & 1 \\ 0 & 0 & 0 \\ 0 & 0 & 0 \end{bmatrix} \qquad (1.1)$$

You should take a moment to think about the fact that you cannot reduce these matrices any further using row operations. For example, in the top left matrix you cannot change the "a" to 0 (if it is not already 0) because the only option would be to multiply the second row by $-\frac{a}{b}$ and add it to the top row, but then the 0 in the second entry of the top row would become $-\frac{a}{b}$, which then cannot be "removed". Likewise, you cannot eliminate the "b" (if it is not already 0) without changing the first entry of the second row to a non-zero number. In other words, another way to view reduced echelon form is that you "cannot make any further progress with row operations".

At this point you may be thinking that one matrix is missing from the above list, which is in reduced echelon form according to the definition:

$$\begin{bmatrix} 0 & 0 & 0 \\ 0 & 0 & 0 \\ 0 & 0 & 0 \end{bmatrix}$$

This matrix is called the zero-matrix of size 3×3. There is a zero-matrix of every size. The $m \times n$ zero-matrix is denoted by $\mathbf{0}_{m \times n}$, but often we simply write $\mathbf{0}$ because the size is clear from the context—as you will see later. It should also be clear that you cannot start with a matrix that is *not* the zero matrix and reduce it to $\mathbf{0}$ with row operations. After all, at some point there would be one remaining row with a non-zero entry—for example, in the second set of 3×3 matrices above (1.1). Having reached that point, what row operation would you use to make the top row all 0s? Therefore, the zero matrix will never be the starting or ending matrix for Gauss-Jordan elimination. The presence of 0-columns (columns with all entries 0) will be discussed in more detail later.

It is a very important fact that the reduced echelon form of a matrix is *unique*. This means that no matter what row operations you use to get the matrix to reduced echelon form, the reduced echelon form is always the same. This is an extremely important theorem that we will prove later (Theorem 2.7). But this theorem is certainly useful. For example, knowing that it is impossible to get a "wrong" or different answer for reduced echelon form, computer scientists are free to tackle the problem of implementing *efficient* routines to obtain reduced echelon form using whatever row operations work the best, without having to worry that different row operations can produce a different answer. (Answers could be incorrect for other reasons, such as rounding errors, but that's a topic for a different course.)

MP 8 *GAUSS-JORDAN ELIMIMATION II*

1.3   Algebraic Operations on Matrices

The first algebraic matrix operation that you will learn is *scalar multiplication*, i.e. multiplying a matrix by a scalar. This is very similar to the row operation **RO1** from Gauss-Jordan elimination, except that the entire matrix is being multiplied by the scalar, not just a single row. That is, if **A** is a matrix and k is a scalar, then the scalar multiple $k\mathbf{A}$ is obtained by multiplying each entry of **A** by k. Written in entry form when $\mathbf{A} = [a_{ij}]$, $k\mathbf{A} = [ka_{ij}]$. We can also write $(k\mathbf{A})_{ij} = ka_{ij}$.

MP 9 *SCALAR PRODUCT*

The next matrix operation you will learn is *matrix addition*. In order to be added, matrices **A** and **B** must be the *same size*. If $\mathbf{A} = [a_{ij}]$ and $\mathbf{B} = [b_{ij}]$ then $\mathbf{A} + \mathbf{B}$ is the matrix $[a_{ij} + b_{ij}]$. We can also write $(\mathbf{A} + \mathbf{B})_{ij} = a_{ij} + b_{ij}$. That is, to add matrices (which must be of the same size), you simply add the corresponding entries.

MP 10 *MATRIX ADDITION*

MP 11 *MATRIX ADDITION II*

As for matrix multiplication, a natural idea would be to simply multiply corresponding entries, analogous with matrix addition. That is, define $(\mathbf{AB})_{ij} = a_{ij}b_{ij}$. This product, known as the *naïve* or *Hadamard product*, is actually *not* the correct definition for the vast majority of applications in science, engineering, and mathematics. While the Hadamard product is superficially simpler than the standard method of matrix multiplication that will be introduced next, it is connected to deeper theorems in linear algebra that are beyond the scope of this course.

The standard method of matrix multiplication is more complicated to define, but it corresponds to the reality of many applications in science and engineering. To begin with, matrices must have *compatible sizes*: **AB** is defined only when **A** has the same number of columns as **B** has rows. Put another way, an $m \times n$ matrix can only be multiplied by an $n \times k$ matrix, and as we will see, the result is an $m \times k$ matrix. You might remember this as the "n's cancelling". This definition creates the situation, which may seem odd to you at first, in which two matrices can be added but not multiplied, and *vice* *versa*. To begin with, matrices $\mathbf{A}$ and $\mathbf{B}$ can be added if and only if they are both $m \times n$ matrices. Now if the $m \times n$ matrix $\mathbf{A}$ and the $m \times n$ matrix $\mathbf{B}$ can be multiplied, it must be true that $m = n$. That is, two matrices, $\mathbf{A}$ and $\mathbf{B}$, can be *both added and multiplied if and only if they are* **square** *matrices of the same size*.

The other possibly disquieting feature of matrix multiplication is that $\mathbf{AB}$ may be defined, but $\mathbf{BA}$ may not be defined. For example, if $\mathbf{A}$ is 2×3 and $\mathbf{B}$ is 3×4. In fact, as you can check, $\mathbf{AB}$ *and* $\mathbf{BA}$ *are both defined if and only if* $\mathbf{A}$ *is an* $m \times n$ matrix and $\mathbf{B}$ is an $n \times m$ matrix.

In order to define matrix multiplication, we will first define *row and column vectors*. You have probably heard the word "vector" in a calculus or physics course, for example, in reference to the velocity vector of a particle in the plane or in space. We will formally define vectors later, but for now you can just think of the word "vector" as describing a special kind of matrix. A *row vector* $\mathbf{R}$ is a $1 \times n$ matrix; that is, $\mathbf{R}$ has exactly one row and n columns. For example: $\mathbf{R} := \begin{bmatrix} 1 & -1 & 5 & 6 \end{bmatrix}$ is a 1×4 row vector. A column vector $\mathbf{C}$ is an $m \times 1$ matrix; for example $\mathbf{C} := \begin{bmatrix} 3 \\ -3 \\ 4 \\ 1 \end{bmatrix}$ is a 4×1 column vector.

According to what was discussed previously, the product $\mathbf{RC}$ is defined and will be a 1×1 matrix. We define $\mathbf{RC}$ by multiplying the first entry of $\mathbf{R}$ by the top entry of $\mathbf{C}$, then adding the product of the second entry of $\mathbf{R}$ times the second entry in $\mathbf{C}$, etc. In this case,

$$\mathbf{RC} = [(1 \cdot 3) + ((-1) \cdot (-3)) + (5 \cdot 4) + (6 \cdot 1)] = [32].$$

If you have had vector calculus, you should recognize that the number 32 is the *dot product* of $\mathbf{R}$ and $\mathbf{C}$ if they are considered as vectors $(1, -1, 5, 6)$ and $(3, -3, 4, 1)$. Whether or not you have seen dot products before is not important at the moment—we will discuss them in more detail later. For now we simply refer to the entry of $\mathbf{RC}$ (in this case the number 32) as the "dot product" $\mathbf{R} \cdot \mathbf{C}$ of $\mathbf{R}$ and $\mathbf{C}$. That is, the dot product is a **scalar**, 32, while the product of these two matrices is the 1×1 matrix [32]. The dot product has many important applications in science, so it should not be a surprise that it is showing up here in a natural way.

MP 12 *DOT PRODUCT*

We now come to the formal definition of the product of an $m \times n$ matrix $\mathbf{A}$ times an $n \times k$ matrix $\mathbf{B}$, which will be an $m \times k$ matrix $\mathbf{C} = \mathbf{AB}$. In order to compute the entry c_{ij} of $\mathbf{C}$, take the dot product of the ith row of $\mathbf{A}$, which we will refer to as $\mathbf{A}_i$ with the jth column of $\mathbf{B}$, which we will refer to as $\mathbf{B}^j$. Note that our requirement on the *size* of these matrices means that we will always be able to carry out these dot products. In fact, this is another way to check the compatibility to multiply matrices. Is a row of the first matrix compatible with the column of the second? Consider for example the product

$$\begin{bmatrix} 0 & 1 \\ 2 & -1 \\ 1 & 2 \end{bmatrix} \begin{bmatrix} 1 & -2 & 0 \\ 4 & 2 & 1 \end{bmatrix}.$$

We are multiplying a 3×2 matrix by a 2×3 matrix, which results in a 3×3 matrix $\mathbf{C} = [c_{ij}]$. To get the entry c_{11}, dot the top row in the left matrix with the first column in the right matrix to obtain $0 \cdot 1 + 1 \cdot 4 = 4$. To get the entry c_{32}, dot the third row in the left matrix with the second column in the right matrix, to get $1(-2) + 2 \cdot 2 = 2$. So we have filled in two entries:

$$\begin{bmatrix} 4 & * & * \\ * & * & * \\ * & 2 & * \end{bmatrix}$$

The rest of the entries in the product can be calculated similarly, and you will have an opportunity to do so shortly. But as you can see, even computing the products of relatively small matrices by hand is tedious, with many chances to make small mistakes. In years long past, when the term "computer" referred to a person who could do tedious calculations relatively quickly and accurately, practicing this basic arithmetic was important for anyone entering a scientific field. Now the term "computer" refers to a machine that can do these computations blindingly fast and with complete accuracy (except for possible rounding errors). No future employer will ask you to compute the product of matrices by hand (or if they do, you probably should find another job because the company will not survive!). In the MPs you will be asked to compute some entries by hand, again just to help cement your understanding of the process, but after that, when doing higher-level problems, the MPs will do the tedious work for you and allow you to focus on the bigger picture.

There are some alternative but equivalent ways to describe matrix multiplication, which we will use in the future. As mentioned above, $\mathbf{A}_i$ denotes the ith row of an $m \times n$ matrix $\mathbf{A}$, and $\mathbf{B}^j$ denotes the jth column of an $n \times k$ matrix $\mathbf{B}$. Then we can describe the components of $\mathbf{AB}$ as follows:

$$(\mathbf{AB})_{ij} = \mathbf{A}_i \cdot \mathbf{B}^j = a_{i1}b_{1j} + a_{i2}b_{2j} + \cdots + a_{in}b_{nj} \tag{1.2}$$

MP 13 *MATRIX PRODUCT*

MP 14 *MATRIX PRODUCT II*

It is sometimes useful to have an expression for the rows and columns of the product of matrices. We will check the following formula:

$$(\mathbf{AB})_i = \mathbf{A}_i \mathbf{B} \tag{1.3}$$

That is, to find the ith row of the matrix $\mathbf{AB}$, take the product of the row $\mathbf{A}_i$ with the matrix $\mathbf{B}$. According to Formula (1.2), the jth entry of the row $(\mathbf{AB})_i$ is $\mathbf{A}_i \cdot \mathbf{B}^j$. On the other hand, the jth entry of the matrix $\mathbf{A}_i\mathbf{B}$ is also $\mathbf{A}_i \cdot \mathbf{B}^j$. Therefore, these two row matrices are the same. Similarly,

$$(\mathbf{AB})^j = \mathbf{AB}^j \qquad (1.4)$$

There is a final important feature of matrix multiplication: Even when the products $\mathbf{AB}$ and $\mathbf{BA}$ both are defined, it is possible that $\mathbf{AB} \neq \mathbf{BA}$. That is, matrix multiplication does not necessarily satisfy the *commutative law*.

MP 15 *COMMUTATIVE LAW*

The non-commutativity of matrix multiplication may seem annoying to students, and yet this fact is simply a reflection of the *reality* that can be represented by matrices. For example, consider the following two non-commuting matrices, which you saw in MP 15. As we will see in more detail later, multiplication by these matrices represents rotations in 3-dimensional space:

$$\mathbf{M}_1 = \begin{bmatrix} 1 & 0 & 0 \\ 0 & 0 & -1 \\ 0 & 1 & 0 \end{bmatrix} \text{ and } \mathbf{M}_2 = \begin{bmatrix} 0 & -1 & 0 \\ 1 & 0 & 0 \\ 0 & 0 & 1 \end{bmatrix} \qquad (1.5)$$

For now you should just accept without complete justification that $\mathbf{M}_1$ can be used via multiplication on the left, to rotate $\frac{\pi}{2}$ radians (90 degrees) in space about the x-axis, which takes z-axis to the y-axis via the product

$$\begin{bmatrix} 1 & 0 & 0 \\ 0 & 0 & -1 \\ 0 & 1 & 0 \end{bmatrix} \begin{bmatrix} x \\ y \\ z \end{bmatrix} = \begin{bmatrix} x \\ -z \\ y \end{bmatrix}$$

Thinking of the vector $\begin{bmatrix} x \\ y \\ z \end{bmatrix}$ as the point (x, y, z), then $(x, y, z) \rightarrow (x, -z, y)$.

That is, $(1, 0, 0) \rightarrow (1, 0, 0)$, $(0, 1, 0) \rightarrow (0, 0, 1)$, and $(0, 0, 1) \rightarrow (0, -1, 0)$. You should convince yourself, using an actual ball if necessary, that this motion fixes the x-axis and rotates the y, z-plane by taking the y-axis to the z-axis and the z-axis to the y-axis. Since $(0, 0, 1) \rightarrow (0, -1, 0)$, the rotation about the x-axis is $\frac{\pi}{2}$ is in the *negative* direction according to the "right hand rule". That is, if you imagine a peanut butter jar around the x-axis with its lid on the right end, this rotation would "screw on the lid" of the jar.

Exercise 1.1 *For the rotation* $\mathbf{M}_2$,

> *1. Do a similar informal analysis of the rotation checking what happens to the three "basis points" $(1, 0, 0)$, $(0, 1, 0)$, and $(0, 0, 1)$ to describe the rotation as a $\frac{\pi}{2}$ radian rotation about the z-axis, including which way the rotation goes.*

2. *Write down the formula for*

$$\mathbf{M}_1\left(\mathbf{M}_2\left(x, y, z\right)\right)$$

as above, i.e. $(x, y, z) \to ?$, *and show where* $\mathbf{M}_1\mathbf{M}_2$ *takes the "basis points"*

3. *Write down the formula for*

$$\mathbf{M}_2\left(\mathbf{M}_1\left(x, y, z\right)\right)$$

and check the formula is different than the one for the first part; then show where $\mathbf{M}_2\mathbf{M}_1$ *takes the "basis points".*

The matrix product $\mathbf{M}_1\mathbf{M}_2$ represents the result of rotating by $\mathbf{M}_2$ first, followed by $\mathbf{M}_1$ and the product $\mathbf{M}_2\mathbf{M}_1$ represents the result of rotating by $\mathbf{M}_1$ first, followed by $\mathbf{M}_2$. You should visualize these two rotations, using an actual ball if needed, to verify geometrically that they do not commute.

1.4 Algebraic Properties of Matrix Operations

By now you have learned the properties of arithmetic of the real numbers, even if you didn't learn (or have forgotten) the names of the rules. We will review these rules now in order to pinpoint which are still true for matrix operations and which are not. We will also take this opportunity to briefly discuss the properties of other scalars. Here are the basic algebraic properties for the real numbers $\mathbb{R}$, which have two operations: multiplication and addition.

Associative Laws: For any real numbers a, b, c, $a + (b + c) = (a + b) + c$ and $a(bc) = (ab)c$.

Commutative Laws: For any real numbers a and b, $a + b = b + a$ and $ab = ba$.

Distributive Law: For any real numbers a, b, c, $a(b + c) = ab + ac$.

Properties of 0 and 1: For any real number a, $a + 0 = a$ and $1a = a$.

Negatives: For any real number a, $-a = (-1)a$ and $a + (-a) = 0$.

Multiplicative Inverses: For any real number $a \neq 0$, the multiplicative inverse of a, denoted by $\frac{1}{a}$ or a^{-1}, satisfies $aa^{-1} = 1$.

From these basic rules, all of the algebraic operations of the real numbers that you have used can be proved as theorems. Among these, the fact that the associative and commutative laws also apply to products and sums of any length. For example, $(a(bc)) d = a(b(cd)) = d(b(ac))$. That is, when working with only a single operation, the order of the operations is not important. That is why one can simply write $abcd$ without parentheses.

The scalars used for matrix algebra generally satisfy all of the above properties, and are referred to as *fields* (for technical reasons one must include that $0 \neq 1$). The real numbers $\mathbb{R}$ are a field, of course, but so are the complex numbers $\mathbb{C}$. If you haven't learned about complex numbers before, don't worry—we are bringing them up now just to make a point and will discuss them later in more detail. All of the basic algebraic rules for matrices are proved using only using the above algebraic rules for fields, and not any other special properties of the real numbers. This will make the transition to matrix algebra using complex matrices very smooth–all of the basics work in exactly the same way! On the other hand, $\mathbb{C}$ has some very basic algebraic features that $\mathbb{R}$ lacks—for example, every polynomial with coefficients in $\mathbb{C}$ has zeros, and can be factored into linear factors. On the other hand, as you well know, the polynomial $x^2 + 1$ has no zeros when x is a real number (this is simply because the polynomial is always positive). But when x is permitted to be a complex number, the complex number i is a solution to this equation; that is, $i^2 = -1$. This is a very important advantage in using $\mathbb{C}$ as the set of scalars, which is a later topic.

Another example that might have seemed like mostly a curiosity in the past, but which now has important relevance to computer science, is the smallest possible field: $\mathbb{F}_2 = \{0, 1\}$. Here multiplication and addition are defined in the usual way, except that of course we must re-define $1 + 1$, since 2 is not in $\mathbb{F}_2$. We only have two options, of course: $1 + 1 = 0$ and $1 + 1 = 1$. As you will verify below, only the first option makes $\mathbb{F}_2$ into a field. This means that, in the language of the rules above, $-1 = 1$. Strange as it seems at first, matrix algebra with scalars in $\mathbb{F}_2$ is used in important applications, including data analysis. The good news is that since $\mathbb{F}_2$ is a field, all of the basic rules for linear algebra hold for matrices in $\mathbb{F}_2$.

Exercise 1.2 *Check that $\mathbb{F}_2$, with these definitions, does satisfy all of the algebraic rules for a field. This just means checking a lot of cases!*

Exercise 1.3 *Check that if you define $1 + 1 = 1$, then $\{0, 1\}$ is not a field. To do this you only need to identify a single field property that is not satisfied.*

Exercise 1.4 *Multiply the following matrices with entries in $\mathbb{F}_2$.*

$$1. \begin{bmatrix} 1 & 1 \\ 1 & 1 \end{bmatrix} \begin{bmatrix} 1 & 1 & 1 \\ 1 & 1 & 1 \end{bmatrix}$$

$$2. \begin{bmatrix} 1 & 0 & 0 \\ 1 & 1 & 0 \\ 1 & 0 & 1 \end{bmatrix} \begin{bmatrix} 1 \\ 1 \\ 0 \end{bmatrix}$$

We also note that computers cannot handle irrational numbers precisely in decimal form because irrationals by definition have infinite, not-repeating decimal expressions. Therefore, strictly speaking, all numerical (as opposed to symbolic) computations done by computers use the rational numbers $\mathbb{Q}$ (i.e. real numbers represented by fractions). $\mathbb{Q}$ is a subset of $\mathbb{R}$ that is closed under addition and multiplication. You know this already—the sum and product of fractions is a fraction. But this means that $\mathbb{Q}$ satisfies all the rules to be a field. There are also deeper mathematical theorems in which it is useful to use $\mathbb{Q}$ rather than $\mathbb{R}$, but these are beyond the scope of this text.

Finally, it is sometimes useful to consider matrices with *integer* scalars $\mathbb{Z} = \{\ldots, -2, -1, 0, 1, 2, \ldots\}$. Note that $\mathbb{Z}$ is definitely not a field—for example, the multiplicative inverse of 2 is $\frac{1}{2}$, which is not an integer! Therefore, one must be careful in that one cannot generally carry out division. "Division", on the other hand, is just another word for multiplying by the multiplicative inverse of a number. For example, $\frac{5}{2}$ is just another way of writing $5 \cdot \frac{1}{2}$ or $5 \cdot 2^{-1}$. As we will see soon, it is not generally possible to "divide" a matrix by another matrix, but there is a notion of multiplicative inverse for matrices. At any rate, the algebraic rules for matrices are generally more complicated than those for scalars. For example, as discussed earlier, it is not always true that two matrices can be multiplied or added. This requires various caveats for our re-statements of the rules for matrix algebra. We start with the rules for matrix addition, which are a bit simpler than those involving matrix multiplication.

Suppose that $\mathbf{A}, \mathbf{B}, \mathbf{C}$ are matrices of the same size. Then

Associative Law for Matrix Addition: $\mathbf{A} + (\mathbf{B} + \mathbf{C}) = (\mathbf{A} + \mathbf{B}) + \mathbf{C}$

Commutative Law for Matrix Addition: $\mathbf{A} + \mathbf{B} = \mathbf{B} + \mathbf{A}$

In effect these rules imply that if one is only adding matrices, then the summation can be done in any order, just like adding real numbers. Since matrix addition is so easy to define, it is easy to see why these laws are true. For example, suppose that $\mathbf{A} = [a_{ij}]$ and $\mathbf{B} = [b_{ij}]$ are matrices of the same size. Then by definition, $(\mathbf{A} + \mathbf{B})_{ij} = a_{ij} + b_{ij}$. But $(\mathbf{B} + \mathbf{A})_{ij} = b_{ij} + a_{ij}$, and by the commutative law (for the scalars), $a_{ij} + b_{ij} = b_{ij} + a_{ij}$. Since $\mathbf{A} + \mathbf{B}$ and $\mathbf{B} + \mathbf{A}$ have the same entries, they are the same matrix. We can write this out more "cleanly" as

$$(\mathbf{A} + \mathbf{B})_{ij} = a_{ij} + b_{ij} = b_{ij} + a_{ij} = (\mathbf{B} + \mathbf{A})_{ij}$$

Let's prove the associative law, letting $\mathbf{C} = [c_{ij}]$. Then using the associative law for scalars,

$$(\mathbf{A} + (\mathbf{B} + \mathbf{C}))_{ij} = a_{ij} + (b_{ij} + c_{ij}) = (a_{ij} + b_{ij}) + c_{ij} = ((\mathbf{A} + \mathbf{B}) + \mathbf{C})_{ij}$$

As for negatives, we define $-\mathbf{A} = [-a_{ij}]$. That is, to *negate* a matrix, we simply negate all of its entries. Doing so is clearly the same as multiplying the matrix by the scalar -1. Term-wise we see

$$(\mathbf{A} + (-\mathbf{A}))_{ij} = a_{ij} + (-a_{ij}) = 0$$

That is,

$$\mathbf{A} + (-\mathbf{A}) = \mathbf{0}$$

(As a reminder, the $\mathbf{0}$ above is **bold**, meaning it represents the zero matrix, not the scalar 0.) As with real numbers, notationally we often write $\mathbf{A} - \mathbf{B}$

rather than $\mathbf{A} + (-\mathbf{B})$. While they may seem almost too simple to write down, we will state the rule for negatives and $\mathbf{0}$, mostly to contrast later with the more complicated rules for multiplicative inverses that you will see below.

Negatives and 0 for Matrices: If $\mathbf{A}$ is any matrix, then $-\mathbf{A} = (-1)\mathbf{A}$, $\mathbf{A} - \mathbf{A} = \mathbf{0}$, and $\mathbf{0} + \mathbf{A} = \mathbf{A} + \mathbf{0} = \mathbf{A}$.

You should take a moment to ponder how much more difficult the proofs will be for rules involving matrix multiplication, due to the complicated definition. The good news (if you want to view it that way) is that, as we have already observed, multiplication is *not commutative*, so that's one less rule to remember! On the other hand, by now you probably use the commutativity of multiplication of real numbers without thinking about it, which means you will have to remain vigilant while doing matrix algebra.

The statement of the associative law for matrix multiplication is complicated by the compatibility condition for multiplication. As a reminder, we can only multiply an $m \times n$ matrix by an $n \times k$ matrix, and the result is an $m \times k$ matrix. Now let's suppose $\mathbf{A}(\mathbf{BC})$ is defined (i.e. the compatibility conditions are satisfied for both $\mathbf{BC}$ and $\mathbf{A}(\mathbf{BC})$). The first thing to wonder about is whether $(\mathbf{AB})\mathbf{C}$ is even defined! The compatibility conditions for $\mathbf{BC}$ mean that $\mathbf{B}$ is $m \times n$ and $\mathbf{C}$ is $n \times k$, so $\mathbf{BC}$ is $m \times k$. This means that $\mathbf{A}$ must be $p \times m$. On the other hand, we see that $\mathbf{AB}$ is defined, and is $p \times n$. But then $(\mathbf{AB})\mathbf{C}$ is also defined. Likewise, one can check that if $(\mathbf{AB})\mathbf{C}$ is defined, then $\mathbf{A}(\mathbf{BC})$ is defined. It is convenient to simplify these two compatibility conditions using the "if and only if" logical construction. We will do so in the formal statement of the associative law for multiplication. So we have proved the first part of the associative law.

Associative Law for Matrix Multiplication: Suppose that $\mathbf{A}, \mathbf{B}, \mathbf{C}$ are matrices. $\mathbf{A}(\mathbf{BC})$ is defined if and only if $(\mathbf{AB})\mathbf{C}$ is defined, and when both are defined, $\mathbf{A}(\mathbf{BC}) = (\mathbf{AB})\mathbf{C}$.

The proof of this statement isn't difficult, but it involves a somewhat tedious double summation to describe each entry in both products. It is great practice using summation notation if you want to give it a shot. On the other hand, a complete proof of the associative law for products of matrices involves an even more tedious proof by induction. Since neither proof gives much insight, I'll omit them. The kind of statement you will see is something like this: $(\mathbf{A}(\mathbf{BC}))\,\mathbf{D}$ is defined if and only if $\mathbf{A}(\mathbf{B}(\mathbf{CD}))$ is defined, and in either case the two products are equal. The difference of course is that there are a lot more ways to write the product of four matrices than of three, but as soon as one is defined, all of the others are also defined, and all are equal. The formal proof of the associative law, even for the real numbers, is actually a quite complicated proof by induction, often only seen in graduate courses in mathematics. In fact, as already mentioned, the proofs of most of the additional basic properties of matrix multiplication are beyond the scope of this course. You will test out the Associative Law in the next MP, but this does not constitute a proof!

MP 16 *ASSOCIATIVE LAW*

The Associative Law does tell us that we can write products without parentheses, i.e. **ABCD**, but you must be careful to realize that we cannot change the *order* of the multiplication since it is not commutative—for example, it could be that $\mathbf{ABCD} \neq \mathbf{ACBD}$.

As with scalars, if **A** is a square matrix, we write $\mathbf{A}^2 = \mathbf{A} \cdot \mathbf{A}$, and more generally the product of **A** with itself n times is written $\mathbf{A}^n$. The associative law tells us we don't have to bother with parentheses, e.g. $\mathbf{A}(\mathbf{AA}) = (\mathbf{AA})\mathbf{A} = \mathbf{A}^3$.

Exercise 1.5 *A matrix* **R** *is called a square root of a matrix* **A** *if* $\mathbf{R}^2 = \mathbf{A}$.

1. *Write out the matrix* $\mathbf{R}^2$, *where*

$$\mathbf{R} = \begin{bmatrix} a & b \\ c & d \end{bmatrix}$$

2. *Find four square roots of* $\begin{bmatrix} 4 & 0 \\ 0 & 4 \end{bmatrix}$.

3. *Find two square roots of* $\begin{bmatrix} 5 & 4 \\ 4 & 5 \end{bmatrix}$. *Hint: Write* $5 = 4 + 1$ *and consider the formula that you wrote above for* $\mathbf{R}^2$.

Exercise 1.6 *As you learned long ago, for real numbers a and b, $(a + b)(a - b) = a^2 - b^2$. Find two 2×2 matrices* **A** *and* **B** *such that*

$$(\mathbf{A} - \mathbf{B})(\mathbf{A} + \mathbf{B}) \neq \mathbf{A}^2 - \mathbf{B}^2$$

Hint: What is the difference between matrix arithmetic and real number arithmetic that allows this to happen? Now go back to an earlier MP or example, and steal a couple of matrices that do the trick.

Distributive Law for Matrices: Suppose that $\mathbf{A}, \mathbf{B}$, and **C** are matrices. Then $\mathbf{A}(\mathbf{B} + \mathbf{C})$ is defined if and only if $\mathbf{AB} + \mathbf{AC}$ is defined, and in this case $\mathbf{A}(\mathbf{B} + \mathbf{C}) = \mathbf{AB} + \mathbf{AC}$. Similarly, $(\mathbf{A} + \mathbf{B})\mathbf{C}$ is defined if and only if $\mathbf{AC} + \mathbf{BC}$ is defined, and in this case $(\mathbf{A} + \mathbf{B})\mathbf{C} = \mathbf{AC} + \mathbf{BC}$.

We had to write both statements above because matrix multiplication is not commutative, so we could not derive one from the other. Similarly, you *cannot* conclude that $\mathbf{A}(\mathbf{B} + \mathbf{C}) = \mathbf{BA} + \mathbf{CA}$.

MP 17 *DISTRIBUTIVE LAW*

Exercise 1.7 *Show the following part of the distributive law for matrices: if* $\mathbf{A}(\mathbf{B} + \mathbf{C})$ *is defined, then* $\mathbf{AB} + \mathbf{BC}$ *is defined. Hint: really this is just a statement about the possible sizes so that the left expression is defined, vs. the sizes so that the right expression is defined.*

The role of 1 for scalars, which is formally called the *multiplicative identity*, is played by a matrix that we saw earlier, called the identity matrix $\mathbf{I}$, which is the $n \times n$ matrix having 1s on the diagonal and 0s elsewhere. In cases when it is necessary to specify the size because it may not be clear from the context, we will write $\mathbf{I}_n$. We will use this notation to state the next property but rarely ever again. The statement $\mathbf{AI} = \mathbf{IA} = \mathbf{A}$ is still acceptable because the size of the identity is determined by the context, namely whether $\mathbf{I}$ is multiplied on the left or right.

The Identity Matrix: If $\mathbf{A}$ is an $m \times n$ matrix, then $\mathbf{AI}_n = \mathbf{I}_m \mathbf{A} = \mathbf{A}$. Why this is true should be clear from the following MP:

MP 18 *IDENTITY MULTIPLICATION*

This finally brings us to the subject of multiplicative inverses, a crucial topic that we will explore further in upcoming sections.

Definition 1.1 *If* $\mathbf{A}$ *is a square matrix, we say that* $\mathbf{A}$ *has an inverse if there is a matrix* $\mathbf{B}$ *such that* $\mathbf{AB} = \mathbf{BA} = \mathbf{I}$. *We also say that* $\mathbf{A}$ *is invertible if it has an inverse.*

We need to check an important statement—that inverses, when they exist, are unique. We state this as a theorem:

Theorem 1.1 *If* $\mathbf{B}$ *and* $\mathbf{C}$ *are inverses of* $\mathbf{A}$, *then* $\mathbf{B} = \mathbf{C}$.

Proof. By definition of inverse, $\mathbf{BA} = \mathbf{I} = \mathbf{CA}$. If we multiply each side of the equation on the right by $\mathbf{B}$ ($\mathbf{C}$ will also work), we obtain

$$(\mathbf{BA})\,\mathbf{B} = (\mathbf{CA})\,\mathbf{B}$$

By the Associative Law this is equivalent to

$$\mathbf{B}\,(\mathbf{AB}) = \mathbf{C}\,(\mathbf{AB})$$

By definition of inverse this is equivalent to

$$\mathbf{BI} = \mathbf{CI}$$

and by the identity property this is equivalent to $\mathbf{B} = \mathbf{C}$. In the future we will simplify how we write equivalent statements using $\Leftrightarrow$. In this particular case we could more compactly write:

$$(\mathbf{BA})\,\mathbf{B} = (\mathbf{CA})\,\mathbf{B} \Leftrightarrow \mathbf{B}\,(\mathbf{AB}) = \mathbf{C}\,(\mathbf{AB}) \Leftrightarrow \mathbf{BI} = \mathbf{CI} \Leftrightarrow \mathbf{B} = \mathbf{C}$$

∎

Since we now know that an inverse of $\mathbf{A}$, if it exists, is unique, we can give it a unique name: $\mathbf{A}^{-1}$. The next theorem is very useful:

Theorem 1.2 *If* **A** *and* **B** *are the same size square matrices and have inverses, then* **AB** *is invertible and* $(\mathbf{AB})^{-1} = \mathbf{B}^{-1}\mathbf{A}^{-1}$.

Proof. By definition we need to check

$$(\mathbf{AB})\left(\mathbf{B}^{-1}\mathbf{A}^{-1}\right) = \mathbf{A}\left(\mathbf{BB}^{-1}\right)\mathbf{A}^{-1} = \mathbf{AIA}^{-1} = \mathbf{AA}^{-1} = \mathbf{I}$$

The other direction is an exercise. ∎

Exercise 1.8 *Check that* $(\mathbf{AB})\left(\mathbf{B}^{-1}\mathbf{A}^{-1}\right) = \mathbf{I}$ *to finish the proof of Theorem 1.2.*

We now restate the definition using this notation:

Matrix Inverses: If a square matrix **A** has an inverse $\mathbf{A}^{-1}$ then $\mathbf{AA}^{-1} = \mathbf{A}^{-1}\mathbf{A} = \mathbf{I}$.

There is a subtle point that deserves mention. At present, according to the definition, we still need to verify *both* $\mathbf{BA} = \mathbf{I}$ and $\mathbf{AB} = \mathbf{I}$, as in the proof of Theorem 1.2, to verify that a "candidate" matrix **B** is actually the inverse of **A**. We will find out later that we need only check one of these two properties, but the proof is perhaps surprisingly out of reach at this point in the text.

MP 19 *INVERSE VERIFICATION*

The last rules involve scalar multiplication. None of them are complicated or unexpected.

Scalar Multiplication of Matrices: For any scalars k, s, and matrices $\mathbf{A}, \mathbf{B}$, $k(s\mathbf{A}) = (ks)\mathbf{A}$, $(k + s)\mathbf{A} = k\mathbf{A} + s\mathbf{A}$, and $k(\mathbf{A} + \mathbf{B}) = k\mathbf{A} + k\mathbf{B}$.

Exercise 1.9 *Show that for any matrix* **A** *and scalars* k, s, $(k + s)\mathbf{A} = k\mathbf{A} + s\mathbf{A}$. *Hint: Use component notation.*

There is one more rule that is also easy to state and verify by direct calculation:

Zero-Matrix Multiplication: For any matrix **A**, $\mathbf{0A} = \mathbf{0}$ and $\mathbf{A0} = \mathbf{0}$ for any zero-matrices so that the products are defined.

1.5   The Trace and the Transpose

One of the simplest basic operations on square matrices is the *trace*: the sum of the diagonal elements. For example

$$tr \begin{bmatrix} 1 & 2 & 3 \\ 5 & 6 & 7 \\ 8 & 9 & 0 \end{bmatrix} = 1 + 6 + 0 = 7$$

The trace is obviously simple to compute, and so there will be no MPs to practice it. At first it may seem strange that the trace means anything at all! As you can quickly check, the trace certainly isn't generally preserved by row operations. Yet the trace is involved in deep mathematical relationships involving matrices (with practical applications), one of which we will see later.

The trace provides a good excuse to introduce summation notation, which you may have already seen in calculus, physics, or computer science courses. Suppose that $\mathbf{A} = [a_{ij}]$. That is,

$$\mathbf{A} = \begin{bmatrix} a_{11} & a_{12} & \cdots & a_{1n} \\ a_{21} & a_{22} & \cdots & a_{2n} \\ \vdots & \vdots & \vdots & \vdots \\ a_{n1} & a_{n2} & \cdots & a_{nn} \end{bmatrix}$$

Then $tr\mathbf{A} = a_{11} + a_{22} + \cdots + a_{nn}$. In summation notation this equation is more compact: $tr\mathbf{A} = \sum_{i=1}^{n} a_{ii}$. The idea of this notation is that the entries a_{ii} are all being added, while i ranges from 1 to n. This form makes it convenient to check the *summation property* of the trace, namely that

$$tr(\mathbf{A} + \mathbf{B}) = tr\mathbf{A} + tr\mathbf{B}$$

This also gives a little practice in adding summations. By the associative and commutative laws for real numbers, you can combine (or split apart) two summations involving the same "counter variable" (in this case i) as long as the bottom and top numbers (the *limits of the sum*) are the same. In this particular case the lower limit of the sum is 1 and the upper limit of the sum is n. Let's show the summation property using our normal notation for the entries of $\mathbf{A}$ and $\mathbf{B}$:

$$tr(\mathbf{A} + \mathbf{B}) = \sum_{i=1}^{n} (a_{ii} + b_{ii}) = \sum_{i=1}^{n} a_{ii} + \sum_{i=1}^{n} b_{ii} = tr\mathbf{A} + tr\mathbf{B}$$

Exercise 1.10 *Use summation notation to show that $tr(k\mathbf{A}) = k\,(tr\mathbf{A})$ for any scalar k and square matrix $\mathbf{A}$.*

We can use summation to write out Equation (1.2) for the components of the product by replacing the numbers that count from 1 to n with a "counter variable" k, since i and j are already used to locate the component:

$$(\mathbf{AB})_{ij} = a_{i1}b_{1j} + a_{i2}b_{2j} + \cdots + a_{in}b_{nj} = \sum_{k=1}^{n} a_{ik}b_{kj} \qquad (1.6)$$

Exercise 1.11 *Show that $tr(\mathbf{AB}) = tr(\mathbf{BA})$. Hint: The diagonal elements of $\mathbf{AB}$ can be written using $i = j$ in Equation (1.6). You should write the entire argument using summation notation, starting with*

$$tr(\mathbf{AB}) = \sum_{i=1}^{n} (\mathbf{AB})_{ii} = \sum_{k=1}^{n} a_{ik}b_{ki} = \cdots$$

We conclude this section with another deceptively simple operation. For any $m \times n$ matrix $\mathbf{A}$, the transpose $\mathbf{A}^T$ is the $n \times m$ matrix produced by exchanging the rows and columns of $\mathbf{A}$.

MP 20 *TRANSPOSE*

The transpose has several properties that we will now list for any matrices $\mathbf{A}$ and $\mathbf{B}$ so that the following operations are defined, and scalar k:

1. $(\mathbf{A}^T)^T = \mathbf{A}$
2. $(k\mathbf{A})^T = k(\mathbf{A}^T)$
3. $(\mathbf{A} + \mathbf{B})^T = \mathbf{A}^T + \mathbf{B}^T$
4. $(\mathbf{AB})^T = \mathbf{B}^T \mathbf{A}^T$

The last statement is not a typo—the transpose of a product is the product of the transposes in reverse order. In fact, this reversal *must* happen due to the compatibility rules for multiplication. That is, if $\mathbf{AB}$ is defined, then $\mathbf{A}$ must be $n \times m$ and $\mathbf{B}$ must be $m \times k$. But then $\mathbf{A}^T$ is $m \times n$ and $\mathbf{B}^T$ is $k \times m$, so the only product that is always defined is $\mathbf{B}^T \mathbf{A}^T$. Let's check the rest of this important statement now. To do so, let's first look at the component form of the transpose. Since rows and columns are being swapped, this means simply reversing i and j for each component. That is,

$$(\mathbf{A}^T)_{ij} = \mathbf{A}_{ji}$$

Now we can compute using the summation form of the product in Equation (1.6). Note that for the first summation below, we are swapping i and j in Equation (1.6) because we are writing out $(\mathbf{AB})_{ji}$ rather than $(\mathbf{AB})_{ij}$. In order to simplify the calculation we will also let

$$a'_{ij} = \left(\mathbf{A}^T\right)_{ij} = a_{ji} \text{ and } b'_{ij} = \left(\mathbf{B}^T\right)_{ij} = b_{ji}$$

We now compute, first using the definition of the transpose, followed by Equation (1.6), followed by the commutative law and Equation (1.6) again at the end:

$$\left((\mathbf{AB})^T\right)_{ij} = (\mathbf{AB})_{ji} = \sum_{k=1}^{n} a_{jk}b_{ki} = \sum_{k=1}^{n} b_{ki}a_{jk} = \sum_{k=1}^{n} b'_{ik}a'_{kj} = \left(\mathbf{B}^T \mathbf{A}^T\right)_{ij}$$

Note that the replacement of b_{ki} by b'_{ik} and a_{jk} by a'_{kj} allows us to use Equation (1.6), since this puts "k in the middle". I cannot overemphasize how important it is for you to work through the calculations in this section very carefully. It is really "just a calculation", but you should be able to write this argument out completely on your own—and should not be satisfied that you "understand" it until you can do that. Calculations like this are extremely common in engineering, physics, and mathematics, as well as many other math-intensive fields.

MP 21 *TRANSPOSE OF PRODUCT*

Exercise 1.12 *Show that if* **A** *and* **B** *are* $m \times n$ *matrices then each of the following products is defined, and is a square matrix:*

$$\mathbf{A}^T\mathbf{B}, \ \mathbf{A}\mathbf{B}^T, \ \mathbf{B}\mathbf{A}^T, \ \mathbf{B}^T\mathbf{A}$$

Symmetric matrices as defined earlier are precisely those matrices **A** such that $\mathbf{A} = \mathbf{A}^T$. A modification of this definition gives you *skew symmetric*: $\mathbf{A} = -\mathbf{A}^T$. Skew symmetric matrices are also sometimes called "anti-symmetric" matrices.

Exercise 1.13 *Prove that a skew symmetric matrix is square, with all diagonal elements equal to* 0.

1.6   Inverses of Matrices

In some ways it is easier to remember the much smaller list of what does *not always* work with matrix multiplication rather than what does. For example, long ago you learned that if a and b are real numbers and $ab = 0$, then it must be true that $a = 0$ or $b = 0$. To see why this is true, suppose that $a \neq 0$. Then we can multiply both sides of the equation by a^{-1} to get $b = a^{-1}0 = 0$. That is, if a is not 0 then b must be 0, and *vice versa*. In the next MP you will see that this same rule is *not always true* for matrix multiplication. In fact, you will multiply two matrices that have *no zeros* in any of their entries, and get the matrix **0**. If $\mathbf{AB} = \mathbf{0}$ and $\mathbf{A}, \mathbf{B} \neq \mathbf{0}$ then **A** is called a *zero divisor* of **B**.

Exercise 1.14 *Show that if* $\mathbb{F}$ *is a field,* $\mathbb{F}$ *has no zero divisors. Hint: Use the same proof as for the real numbers in the previous paragraph, checking carefully which axioms you are using.*

MP 22 *ZERO DIVISOR*

Yet another difference involving matrix multiplication is that there is no "cancellation law" for matrices. Recall that if a, b and c are real numbers with $a \neq 0$ and $ab = ac$ then $b = c$. This is true in any field because when $a \neq 0$ you can simply multiply both sides of the equation $ab = ac$ by a^{-1} to get $b = c$. In the next MP you will see that this rule also can fail for matrix multiplication.

MP 23 *NO CANCELLATION*

In each of the above cases, the proofs for these two rules for the real numbers involved multiplying by a^{-1}, and a^{-1} exists for real numbers (or scalars in any field) as long as $a \neq 0$. Along these lines, suppose that a matrix $\mathbf{A}$ *does* have an inverse $\mathbf{A}^{-1}$. Then using the rules for matrix multiplication we can carry out the exact same arguments as for a field, with the identity matrix $\mathbf{I}$ playing the role of the number 1: If $\mathbf{AB} = \mathbf{0}$ then

$$\mathbf{B} = \mathbf{IB} = \left(\mathbf{A}^{-1}\mathbf{A}\right)\mathbf{B} = \mathbf{A}^{-1}\left(\mathbf{AB}\right) = \mathbf{A}^{-1}\mathbf{0} = \mathbf{0}$$

Exercise 1.15 *For each "=" in the above equation, specify which rule of matrix algebra that you are using. For example, the last "=" is the property* **Zero-Matrix Multiplication.**

Exercise 1.16 *Prove the following partial cancellation law: If* $\mathbf{BA} = \mathbf{CA}$ *and* $\mathbf{A}$ *has an inverse* $\mathbf{A}^{-1}$ *then* $\mathbf{B} = \mathbf{C}$. *Be careful not to inadvertently use the commutative law, which does not always hold for matrix multiplication!*

Although you may be tiring of all the warnings about non-commutativity, I will still point out that if $\mathbf{A}^{-1}$ exists, then from the equation $\mathbf{AB} = \mathbf{CA}$, the only conclusions one may draw are $\mathbf{B} = \mathbf{A}^{-1}\mathbf{CA}$ and $\mathbf{C} = \mathbf{ABA}^{-1}$. There are many examples in which, for example, $\mathbf{A}^{-1}\mathbf{CA} \neq \mathbf{C}$.

All of this brings us to the following important question: how do we know when a square matrix has an inverse, and if it has an inverse, how do we find it? In MP 22 you verified that $\begin{bmatrix} 2 & 4 \\ 6 & 12 \end{bmatrix} \begin{bmatrix} 2 & -4 \\ -1 & 2 \end{bmatrix} = \mathbf{0}$. This means that neither of these two matrices can have an inverse, because if it did, then we could multiply both sides by the inverse and conclude that the other matrix must be $\mathbf{0}$. So there are definitely non-zero square matrices that do not have inverses—including each of these two.

The next process, which we will justify later, is a variation of Gauss-Jordan elimination. It will always tell you whether or not a square matrix has an inverse—and it will find the inverse if it has one. By Theorem 1.1, inverses are unique when they exist, so for correctness (but not necessarily efficiency), it doesn't matter how we find them. In fact, using Gauss-Jordan elimination is one of the most efficient methods. Here's the process, which you will practice in MP 24. Given a square matrix $\mathbf{A}$, we will first create the *augmented matrix* with the identity matrix $\mathbf{I}$. This means that we create a new matrix whose entries consist of $\mathbf{A}$ on the left and $\mathbf{I}$ on the right. We then use row operations on this larger matrix until we reduce the left side of the matrix to reduced echelon form. If that form is the identity matrix, then $\mathbf{A}$ has an inverse, and the inverse is the final matrix on the right side. If the reduced echelon form has even a single row of zeros, then $\mathbf{A}$ has no inverse. The setup is as follows for the matrix $\mathbf{A} = \begin{bmatrix} 1 & 2 \\ 3 & 4 \end{bmatrix}$. We first create the augmented matrix with the identity matrix on the right. This matrix has a column of vertical dashes to indicate that it is augmented, meaning that it is made up of two matrices that we want to keep track of while we apply row operations to the entire matrix. So the augmented matrix is

$$\begin{bmatrix} 1 & 2 & | & 1 & 0 \\ 3 & 4 & | & 0 & 1 \end{bmatrix}$$

When we have reduced the left side to reduced echelon form, *doing the same operations on both sides of the augmented matrix,* the augmented matrix becomes

$$\begin{bmatrix} 1 & 0 & | & -4 & 3 \\ 0 & 1 & | & 2 & -1 \end{bmatrix}$$

Since the left side of the augmented matrix is **I**, the original matrix has an inverse, and that inverse is the right side: $\begin{bmatrix} -4 & 3 \\ 2 & -1 \end{bmatrix}$. You can check that the product of these two matrices, in either order, is **I**. If we try the same thing with the matrix $\begin{bmatrix} 2 & -4 \\ -1 & 2 \end{bmatrix}$ that we considered earlier and proved has no inverse, we get the augmented matrix

$$\begin{bmatrix} 1 & 0 & | & \frac{1}{2} & 0 \\ 0 & 0 & | & \frac{1}{2} & 1 \end{bmatrix}$$

Sure enough, the left side is not the identity matrix, verifying that the original matrix has no inverse. Determining that a matrix does *not* have an inverse does not always involve reducing entirely to reduced echelon form on the left. Strictly speaking, you can stop as soon as a row of 0s appears on the left. There are other "early signs" that a matrix doesn't have an inverse—for example, if at some point there are two different rows that are identical, which quickly leads to a row of all zeros. But in the next MP, in order to have a consistent "stopping point", you will be required to always reduce the left side to its unique reduced echelon form before drawing conclusions about whether or not the matrix has an inverse. Nonetheless—pay attention and try to see that the matrix has no inverse at the earliest opportunity!

MP 24 *INVERSES*

There are many ways to calculate inverses of matrices, some more efficient, under some circumstances, than others. Many of these alternate methods were developed in the 19th century (or sometimes earlier!) to assist scientists (and their assistants) who of necessity were doing computations with smaller matrices. There are entire courses about computational methods in linear algebra, but this course is not intended to teach students about efficient algorithms to be coded—something that many students will never do. Rather, the common ground for all introductory linear algebra students is *conceptual*: what are the mathematical processes, what do they signify, and how are they applied?

A square matrix that has an inverse is also referred to as *non-singular*. Likewise, a non-invertible matrix is also called *singular*. We conclude this section with a few general facts about inverses, which we will state as a theorem:

Theorem 1.3 *Let* **A** *and* **B** *be square matrices.*

1. *If* **A** *is invertible then so is* $\mathbf{A}^{-1}$, *and* $\left(\mathbf{A}^{-1}\right)^{-1} = \mathbf{A}$.

2. *If* $k \neq 0$ *is a scalar and* **A** *is invertible, then so is* $k\mathbf{A}$, *and* $(k\mathbf{A})^{-1} = \frac{1}{k}\mathbf{A}^{-1}$.

3. *If* **A** *is invertible then so is* $\mathbf{A}^T$, *and* $\left(\mathbf{A}^T\right)^{-1} = (\mathbf{A}^{-1})^T$.

Proof. At this point, the only way that we have to prove that a matrix is invertible is the definition. That is, if we want to show that **D** is invertible and has inverse **C**, the only strategy is to prove that $\mathbf{CD} = \mathbf{DC} = \mathbf{I}$. This is easiest when we already have a "candidate" matrix **C** for the inverse. Then it is just checking two matrix products, as we did in the proof of Theorem 1.2. The proof of the first part is simple but perhaps tricky: the two equations $\mathbf{AA}^{-1} = \mathbf{I}$ and $\mathbf{A}^{-1}\mathbf{A} = \mathbf{I}$ that show that $\mathbf{A}^{-1}$ is the inverse of **A** are the exact same equations to show that **A** is the inverse of $\mathbf{A}^{-1}$. In fact, it may seem as though we aren't doing anything. But, roughly speaking, we did need to show that **A** "acts like" the inverse of $\mathbf{A}^{-1}$—it is simply the case that the required equations are exactly the same. We will check one of the equations for the second part and leave the last part as an exercise.

$$(k\mathbf{A})\left(\frac{1}{k}\mathbf{A}^{-1}\right) = \left(k\frac{1}{k}\right)(\mathbf{AA}^{-1}) = 1\mathbf{I} = \mathbf{I}$$

■

Exercise 1.17 *Prove the third statement of the above theorem. Hint: By uniqueness, you need only show that* $\mathbf{A}^T(\mathbf{A}^{-1})^T = \mathbf{I}$, *using properties of the transpose.*

2

Systems of Linear Equations and Matrices

In this chapter you will learn about systems of linear equations and how they can be reinterpreted as matrix equations. Matrix algebra and Gauss-Jordan elimination can be used to solve these systems and express their solutions in a convenient way. This chapter also includes some basic applications of linear systems.

2.1 Matrix Form of Systems

One of the most basic equations is the equation of a line in the plane, which you have learned has the form $y = mx + b$, where m is the slope of the line and $(0, b)$ is the y-intercept—the point at which the line intersects the y-axis (except when the line is vertical and doesn't intercept the y-axis). This equation is among the most simple of all equations, and models many of the most basic relationships in the universe, some of which the student will have seen in elementary precalculus and calculus, as well as basic science courses. Even when a relationship is not modeled on linear equations but rather obeys some kind of *differentiable* relationship, linear equations may be used as approximate models for the relationship. Throughout calculus, students often use "hidden" linear algebra—for example, when determining the equation of the tangent line or plane to the graph of a differentiable function. Computing partial derivatives is actually computing the entries in a matrix called the Jacobian matrix.

The equation $y = mx + b$ can be rewritten in the form $-mx + y = b$. In this form, the equation has the general form of a *linear equation* in two variables, that is, an equation of the form $ax + by = c$, where a and b are real numbers called *coefficients*, c is a real number generally referred to as the *constant term*, and x and y are variables. In this form the case of a vertical line is clear: $b = 0$ and $a \neq 0$, so the equation is $ax = c$.

One can use any letters for coefficients, constant, and variables, but there are certain conventions that help those learning and using mathematics to

DOI: 10.1201/9781003674368-2

keep track of which are variables and which are coefficients. Very often u, v, r, s, t, x, y, z are used for variables, while letters near the beginning of the alphabet are used for coefficients and constant terms. For example, a linear equation in the variables x, y, z has a general form $ax + by + cz = d$. A specific equation would be $2x + 4y - 2z = 4$. There are certain traditions involving coefficients ± 1 and 0. The coefficient ± 1 is usually not written, and the entire term involving a coefficient 0 is usually omitted, as we did above in the case $ax = c$. For example, $0x - 1y + 3z = 4$ would normally be written $-y + 3z = 4$. It is important to remember that in this case this is *still* a linear equation in variables x, y, z–something that can be determined from the context of the problem.

Linear equations may have many variables–in modern applications a linear equation could even have hundreds of variables. At some point it becomes impractical to use different letters, and instead one uses the same letter with subscripts. So a linear equation in six variables generally is of the form

$$a_1 x_1 + a_2 x_2 + a_3 x_3 + a_4 x_4 + a_5 x_5 + a_6 x_6 = c$$

A specific example would be

$$3x_1 - 3x_2 + x_3 - x_5 + 6x_6 = 0$$

Note that there is a "missing" x_4 term, which just means that the coefficient $a_4 = 0$. In this way, use of variables with subscripts can help clarify the number of variables when some of the coefficients are 0. Note also that in this case $a_3 = 1$ is not written, and $a_5 = -1$ is simply written with a minus sign. That is, the above expression is simply "short-hand" for

$$3x_1 + (-3)x_2 + 1x_3 + 0x_4 + (-1)x_5 + 6x_6 = 0$$

You have certainly encountered *non-linear* equations, such as $y = x^2$, $y + 3xz^3 = 5$ or $\sin \theta = 4$.

You are certainly familiar with the real numbers $\mathbb{R}$, which we regard as "one-dimensional". (We will discuss the idea of dimension in much greater detail later.) The two-dimensional space $\mathbb{R}^2$ (also known as *the plane*) consists of ordered pairs (x, y) of real numbers. The three dimensional space $\mathbb{R}^3$ consists of ordered triples (x, y, z). Likewise $\mathbb{R}^4$ consists of quadruples (x, y, z, w) or (x_1, x_2, x_3, x_4). But there is no reason to stop with dimension four, and no reason why (as some students sometimes come to believe from popular works about relativity theory) the fourth dimension must be "time". Variables simply represent possible quantities, and equations represent relationships among the variables. For example, for a probe in the ocean, one might use three variables to describe the location of the probe (including depth) and variables to indicate temperature and pressure. Each data point for the probe has five variables, which we might denote by (x, y, z, t, p), and this quintuple lies in the five-dimensional space $\mathbb{R}^5$.

A collection of linear equations involving the same variables is called a *system of linear equations.* Here is a system of linear equations in three variables, x, y, and z:

$$3x - 4y + z = 2$$
$$x + y - z = 2$$
$$y + 3z = -3 \qquad (2.1)$$
$$x - 2z = 3$$

Note that there is a missing "x" term in the third equation. This is because the term in question is $0x$ and the term is not written. For the same reason there is no "y" term in the last equation, but all of these equations are considered as equations in x, y, and z. A *solution* to this system (if it exists) consists of particular values of x, y, and z so that *all* of the equations are satisfied. In this example, $x = 1$, $y = 0$, and $z = -1$ is a solution to this system–as you can check just by plugging these values into the left side of each equation and checking that the sum is equal to the right side. *Checking* that a set of values is a solution to a system of equations is easy. *Finding* solutions is more difficult.

Not every system of linear equations has a solution, and some systems of linear equations have infinitely many solutions. For example, consider the following system of two linear equations:

$$x - y = 6$$
$$2x - 2y = 4$$

From elementary geometry, one sees that these equations represent two lines in the plane with slope 1 but with y-intercept -6 and -2. That is, these two lines are parallel but not the same line. Since they are parallel, they have no intersection point. An intersection point is a point (x, y) that lies on both lines, which means that those values of x and y are solutions to both equations. Since the lines do not intersect, there are no solutions to this system. If we change the constant on the right side of the second equation from 4 to 12, we get the following system:

$$x - y = 6$$
$$2x - 2y = 12$$

In this new system, both equations represent the same line, which can be seen by dividing both sides of the second equation by 2. Since the equations represent the same line, there are infinitely many ordered pairs (x, y) on "both lines" (which are actually the same line), so the system has infinitely many solutions. In fact, for any choice of x we may solve for y to get a solution. For example, for $x = 0$, $y = -6$ is a solution.

Now replace the second equation by $x - 2y = 4$ to get the system

$$x - y = 6$$
$$x - 2y = 4$$

The line represented by the second equation has a different slope from the line for the first equation. As you may recall from elementary geometry, these two lines meet in a single point. That is, there is exactly one solution to this new system, and as learned in precalculus, we may actually solve for that unique system using various methods. For example, one can subtract the top equation from the bottom one, to obtain a new second equation $-y = -2$ or $y = 2$. Substituting for y we obtain $x - 2 = 6$ or $x = 8$. One can then check that $x = 8$, $y = 2$ is a solution to the system. This strategy is easy to carry out for two equations with two variables, but for larger systems with many variables, one needs a more practical and systematic way to solve linear systems, with a clear algorithm that can be coded to allow a computer to do the work.

The first important observation is that the names of the variables don't matter—all that matters are the coefficients and constant terms. Therefore the very first step in solving systems of linear equations is to get rid of the variables altogether and put the coefficients into an *augmented matrix*. As you saw in the process for finding inverses of matrices (if they exist), an augmented matrix consists of two matrices put together with "separators" between them–usually vertical bars or dots. Consider the system considered earlier:

$$\begin{aligned} 3x - 4y + z &= 2 \\ x + y - z &= 2 \\ y + 3z &= -3 \\ x - 2z &= 3 \end{aligned} \tag{2.2}$$

We will put all of the coefficients (including the 0s and 1s) into a 4×3 matrix (because there are four equations and three variables), then augment this matrix by adding a fourth column with the constants on the right sides of the equations:

$$\left[\begin{array}{ccc|c} 3 & -4 & 1 & 2 \\ 1 & 1 & -1 & 2 \\ 0 & 1 & 3 & -3 \\ 1 & 0 & -2 & 3 \end{array} \right]$$

There is nothing complicated here. Creating the augmented matrix for a system just means copying the coefficients and constants for each equation into a row in the matrix. An augmented matrix contains exactly the same information as a system of linear equations. Given an augmented matrix you can create the corresponding system of linear equations simply by choosing variable names and using the entries in the matrix as coefficients and constants for the equations in the system. The general form of a system of m linear equations with variables $x_1, \ldots, x_n$ looks like this:

$$\begin{aligned} a_{11}x_1 + a_{12}x_2 + \cdots + a_{1n}x_n &= b_1 \\ a_{21}x_1 + a_{22}x_2 + \cdots + a_{2n}x_n &= b_2 \\ &\vdots \\ a_{m1}x_1 + a_{m2}x_2 + \cdots + a_{mn}x_n &= b_m \end{aligned}$$

The matrix

$$A = \begin{bmatrix} a_{11} & a_{12} & \cdots & a_{1n} \\ a_{21} & a_{22} & \cdots & a_{1n} \\ \vdots & \vdots & \vdots & \vdots \\ a_{m1} & a_{m2} & \cdots & a_{mn} \end{bmatrix} = [a_{ij}]$$

is called the *coefficient matrix* of the system, and the augmented matrix

$$\begin{bmatrix} a_{11} & a_{12} & \cdots & a_{1n} & | & b_1 \\ a_{21} & a_{22} & \cdots & a_{1n} & | & b_2 \\ \vdots & \vdots & \vdots & \vdots & | & \vdots \\ a_{m1} & a_{m2} & \cdots & a_{mn} & | & b_m \end{bmatrix}$$

is called the *matrix of the system* or more briefly the *system matrix*.

MP 25 *MATRIX FROM A SYSTEM*

MP 26 *SYSTEM FROM A MATRIX*

2.2   Solving Systems Using Gauss-Jordan Elimination

Gauss-Jordan elimination provides an easily implementable method to solve systems of linear equations. Take the augmented matrix for the system, and use row operations to reduce the *coefficient matrix* to reduced echelon form. For brevity we will refer to this as the *reduced form of the augmented matrix*. You should not reduce the entire matrix to reduced echelon form–just the coefficient matrix on the left. One possible outcome is for the coefficient matrix to reduce to the identity matrix. When this happens, you can immediately read off the unique solution to the system. For example, suppose that we have reduced the augmented matrix to the following:

$$\begin{bmatrix} 1 & 0 & 0 & | & 4 \\ 0 & 1 & 0 & | & 0 \\ 0 & 0 & 1 & | & -3 \end{bmatrix}$$

If we write down the system corresponding to this augmented matrix, we get

$$x = 4$$
$$y = 0$$
$$z = -3$$

In this case, there is one and only one solution specified by these equations; it is unique.

MP 27 *SOLVING SYSTEMS UNIQUE*

How can we be sure that the new system of equations has the same solutions as the original when we have done Gauss-Jordan elimination? Really
the question is this: do the elementary row operations change the solutions
of the corresponding system? Let's look at them one at a time. First, multiplying a row by a non-zero scalar is the same is multiplying both sides of
the corresponding equation by that scalar. But as you have known since your
first high school algebra course, multiplying both sides of an equation by a
non-zero number does not change the solutions to that equation. The reason
you aren't permitted to multiply by 0 is that when you multiply both sides of
any equation by 0, you are replacing that equation by $0 = 0$, which is always
true, so you have basically removed an equation from the system. This can
clearly change the solutions. Obviously, swapping two rows doesn't change
the equations at all–just the order in which they have been written down.
Similarly, adding one equation to another doesn't change the solution since
you are adding equal quantities to both sides of the equation. And finally, replacing one equation by its sum with a scalar multiple of another equation is
the combination of two operations–scalar multiplication of both sides followed
by adding the equations–which therefore doesn't change the solutions. This
paragraph was a proof of the following theorem:

Theorem 2.1 *The row operations in Gauss-Jordan elimination do not
change the solutions of the corresponding system of equations.*

As you know, Gauss-Jordan elimination does not always end with the identity matrix; reduced echelon form may have one or more rows of zeros. With
an augmented matrix, there is also the right column to consider. When starting with a square coefficient matrix (same number of equations and variables,
there are three possible outcomes for an augmented matrix in reduced form:

1. The reduced coefficient matrix has at least one row of 0s with a *non-
   zero* number b in the final entry of that row. In this case, the equation corresponding to that row is $0 = b$, which has **no solutions**
   because $b \neq 0$. A system without solutions is called *inconsistent*.

2. The system is consistent and every column of the reduced coefficient
   matrix has a leading entry. In this case, there is a **unique solution**
   that may simply be read off from the matrix. You saw some of these
   already in MP 27.

3. The system is consistent and there is at least one column with no
   leading entry. As we will detail below, this means that there are
   infinitely many solutions.

The next MP allows you to practice the first outcome. This should always
be checked for first, because if the system is inconsistent, there are no solutions
and that's the end of the story.

MP 28 *SOLVING SYSTEMS NONE*

The third possible outcome–infinitely many solutions–is a bit more complex. It will be important not only to know that there are infinitely many solutions, but to be able to describe them all. It will take some time to reach that point. In the meantime, it is useful to consider the reduced echelon forms for 3×3 coefficient matrices when Gauss-Jordan elimination does not lead to the identity matrix.

Referring back to the list of matrices (1.1), we have the following possible outcomes for the reduced form of a 3×4 augmented matrix.

$$\mathbf{O}_U = \begin{bmatrix} 1 & 0 & 0 & | & b_1 \\ 0 & 1 & 0 & | & b_2 \\ 0 & 0 & 1 & | & b_3 \end{bmatrix}$$

As you know, this option has a unique solution, $x_1 = b_1$, $x_2 = b_2$ and $x_3 = b_3$.

$$\mathbf{O}_1 = \begin{bmatrix} 1 & 0 & a & | & b_1 \\ 0 & 1 & b & | & b_2 \\ 0 & 0 & 0 & | & b_3 \end{bmatrix} \quad \mathbf{O}_2 = \begin{bmatrix} 1 & a & 0 & | & b_1 \\ 0 & 0 & 1 & | & b_2 \\ 0 & 0 & 0 & | & b_3 \end{bmatrix} \quad \mathbf{O}_3 = \begin{bmatrix} 0 & 1 & 0 & | & b_1 \\ 0 & 0 & 1 & | & b_2 \\ 0 & 0 & 0 & | & b_3 \end{bmatrix}$$

In these cases, if $b_3 \neq 0$ then there are no solutions. Otherwise, the system is consistent, and we have the following possibilities, setting $b_3 = 0$:

$$\mathbf{O}_4 = \begin{bmatrix} 1 & 0 & a & | & b_1 \\ 0 & 1 & b & | & b_2 \\ 0 & 0 & 0 & | & 0 \end{bmatrix} \quad \mathbf{O}_5 = \begin{bmatrix} 1 & a & 0 & | & b_1 \\ 0 & 0 & 1 & | & b_2 \\ 0 & 0 & 0 & | & 0 \end{bmatrix} \quad \mathbf{O}_6 = \begin{bmatrix} 0 & 1 & 0 & | & b_1 \\ 0 & 0 & 1 & | & b_2 \\ 0 & 0 & 0 & | & 0 \end{bmatrix}$$

$$(2.3)$$

The final outcome is

$$\mathbf{O}_7 = \begin{bmatrix} 1 & a & b & | & b_1 \\ 0 & 0 & 0 & | & b_2 \\ 0 & 0 & 0 & | & b_3 \end{bmatrix}$$

In this case there is no solution if b_2 or b_3 is non-zero, and if both are 0 there is a single equation $x_1 + ax_2 + bx_3 = 0$.

In all of these cases except $\mathbf{O}_U$, there are columns with no leading entries— for example, in $\mathbf{O}_1$ the third column has no leading entries. This means that there is no equation that involves z by itself (without the other variables). In fact, here are the equations: $x + az = b_1$ and $y + bz = b_2$. We are "free" to choose any value for z and then solve the two other equations for x and y to get a solution. For this reason, z is called a *free variable*. More generally:

Definition 2.1 *If a matrix is in reduced echelon form, a column with no leading entries is called a free column; if the matrix is the coefficient matrix for a system of linear equations, then the variable corresponding to a free column is called a free variable. Columns with a leading entry are called leading columns, and the corresponding variables are called leading variables.*

Exercise 2.1 *For each of the matrices* $\mathbf{O}_U$ *and* $\mathbf{O}_i$ *above, identify the free columns, if any.*

Several of the outcomes above involve columns of 0s, and it is important to see how this can (and can't!) happen.

Theorem 2.2 *If a matrix* $\mathbf{A}$ *has a column* $\mathbf{A}^j$ *that is not all 0s then the* j^{th} *column in reduced echelon form is also not all 0s.*

Proof. There is no row operation that can produce a column of 0s from one that has at least one non-zero term: Multiplying a row by a non-zero scalar cannot change a non-zero entry to zero. Likewise, swapping rows doesn't change any non-zero entries in any column to 0. Using the second row operation, the only way to change a non-zero entry b_{ij} to 0 is to add a non-zero scalar multiple of another row $\mathbf{B}_k$ to $\mathbf{B}_i$. This step also requires that $b_{kj} \neq 0$. That is, changing b_{ij} to 0 requires another entry b_{kj} in the same column to be non-zero. Put another way, no row operation can be used to remove the "last" non-zero entry from a column in the reduction to reduced echelon form. ∎

When creating the coefficient matrix for a system of linear equations, it is very often true that *every variable will be used in some equation*, which means that every column will have at least one non-zero entry in both the coefficient matrix and the reduced echelon form. If there is no equation involving a variable, then it is "free" in the most simple way: there is no equation involving it, so the variable plays no role in the immediate problem. But in some contexts it is possible for a variable to play a role in some systems, but not in others. Therefore, we will still allow the possibility of columns of 0s in $\mathbf{A}$ and its reduced echelon form.

It is now useful to write down carefully what a *reduced form augmented matrix* $\mathbf{R}$ looks like in completely general form. There are certain conventions in mathematical notation that you will get used to with a little practice. For example, sometimes things are depicted that may not exist! In the depiction below, you must keep in mind that there might be 0-columns on the left as shown—see matrix $\mathbf{O}_3$ above—or there might not be, as in $\mathbf{O}_5$. There might not be any 0-rows on the bottom, as in matrix $\mathbf{O}_U$, or there might be, as in all the rest. The other thing to notice is that the entries in free columns (except for those in the bottom 0-rows) are depicted as "$*$". In the depiction below, the column after the first 1 at the top is depicted as a free column, because that is a possibility–as in $\mathbf{O}_5$. But it is also possible that this column is not a free column, as in $\mathbf{O}_4$. It would be extremely cumbersome to write down different depictions involving all of these possibilities for a general matrix, so you should learn to have the "notation flexibility" to see all the options in a single matrix. If this is confusing to you at the moment, it would be helpful to go through all of the matrices $\mathbf{O}_i$ and $\mathbf{O}_U$ and reconcile each one with this depiction. For example, for $\mathbf{O}_U$, there are no 0-columns or 0-rows in the coefficient matrix, and no free columns. In other words, in this case one should ignore all the 0-rows and 0-columns in the depiction of $\mathbf{R}$ below. The main point is that this depiction captures the essence of reduced echelon form of the coefficient matrix: that each leading entry is to the right of those above it, and each leading column has all 0s except for the leading entry.

$$
\begin{bmatrix}
0 & \cdots & 1 & * & \cdots & 0 & * & \cdots & 0 & * & \cdots & | & b_1 \\
0 & \cdots & 0 & 0 & \cdots & 1 & * & \cdots & 0 & * & \cdots & | & b_2 \\
0 & \cdots & 0 & 0 & \cdots & 0 & 0 & \cdots & 1 & * & \cdots & | & b_3 \\
\vdots & \vdots & \vdots & \vdots & \vdots & \vdots & \vdots & \vdots & \vdots & \vdots & \vdots & | & \vdots \\
0 & \cdots & 0 & 0 & \cdots & 0 & 0 & \cdots & 0 & 0 & \cdots & | & b_j \\
\vdots & \vdots & \vdots & \vdots & \vdots & \vdots & \vdots & \vdots & \vdots & \vdots & \vdots & | & \vdots \\
0 & \cdots & 0 & 0 & \cdots & 0 & 0 & \cdots & 0 & 0 & \cdots & | & b_m
\end{bmatrix}
\tag{2.4}
$$

Notation 1 *In what follows, it is necessary for subscripts to involve expressions, so we will tweak the notation a little–although in a standard way that you will see elsewhere. For example, for the entry of* **R** *in row k and column $n + 1$, it is slightly problematic to write the subscript. The obvious options are a bit confusing: r_{kn+1} or $r_{k(n+1)}$. Therefore we insert a comma, writing $r_{k,n+1}$. We will also be using double subscripts below, which you may have seen in calculus concerning subsequences. If you haven't seen this notation before, it may take some getting used to. You will learn this notation best by reading every symbol carefully in the calculations below to be sure you understand what it means.*

Returning to the reduced form matrix **R** (2.4), in each non-zero row $\mathbf{R}_i$ there is a leading entry $r_{ij} = 1$. For this row we may solve for the variable x_j, and since all entries preceding r_{ij} in that row are 0, this expression only involves coefficients r_{ik} with $k > j$. That is,

$$
x_j = b_i - r_{i,j+1} x_{j+1} - \cdots - r_{in} x_n
$$

For example, in the first matrix in (2.2), with $b_1 = b_2 = 0$ we have

$$
x_1 = -a x_3 \tag{2.5}
$$
$$
x_2 = -b x_3
$$

Returning to the row $\mathbf{R}_i$, any coefficient r_{ik} with $k > j$ that lies in a *leading* column must be 0, because leading columns contain only 0s except for the leading entry. Again, you see this in (2.5)—the only possibly non-zero entries come from the third column, which is free. That is, we can write every leading variable as

$$
x_j = b_i - r_{iv} x_v - \cdots - r_{ir} x_r
$$

where the variables $x_v, \ldots, x_r$ are the free variables with $v, \ldots, r > k$. Now for *any* choice $x_v := s_v, \ldots, x_r := s_r$ for the free variables $x_v, \ldots, x_r$ we have a solution

$$
s_j = b_i - r_{iv} s_v - \cdots - r_{ir} s_r \tag{2.6}
$$

We will simplify these expressions in the next theorem using double subscripts. If you are not proficient with double subscripts you should read the statement very carefully and reconcile Equation (2.7) with Equation (2.6) and the matrix (2.4). We summarize this discussion as a theorem:

Theorem 2.3 *If* $\mathbf{R}$ *is in reduced echelon form with free variables* $x_{f_1}, \ldots, x_{f_r}$, *then the solutions to the homogeneous system* $\mathbf{RX} = \mathbf{0}$ *consist of making any choices* $x_{f_1} := s_{f_1}, \ldots, x_{f_r} := s_{f_r}$ *for the free variables, together with the uniquely determined numbers*

$$s_j = \sum_{f_m > j} -r_{if_m} s_{f_m} \tag{2.7}$$

for the leading variables when i is the row that contains the leading entry for x_j. In particular, the system has infinitely many solutions if and only if $\mathbf{R}$ has at least one free column.

For a square matrix, free variables must occur whenever there is a row of 0s across the augmented matrix. To help illustrate Theorem 2.3, I will list the possible reduced forms for a 3×4 augmented matrix for a consistent system in which every variable is in some equation–equivalently there are no 0-columns. (See the matrix (2.4) for the general form.) All of these outcomes except for the first correspond to systems with infinitely many solutions.

$$\begin{bmatrix} 1 & 0 & 0 & | & b_1 \\ 0 & 1 & 0 & | & b_2 \\ 0 & 0 & 1 & | & b_3 \end{bmatrix} \begin{bmatrix} 1 & 0 & * & | & b_1 \\ 0 & 1 & * & | & b_2 \\ 0 & 0 & 0 & | & 0 \end{bmatrix} \begin{bmatrix} 1 & * & 0 & | & b_1 \\ 0 & 0 & 1 & | & b_2 \\ 0 & 0 & 0 & | & 0 \end{bmatrix}$$

$$\begin{bmatrix} 1 & 0 & * & | & b_1 \\ 0 & 1 & * & | & b_2 \\ 0 & 0 & 0 & | & 0 \end{bmatrix} \begin{bmatrix} 1 & * & 0 & | & b_1 \\ 0 & 0 & 1 & | & b_2 \\ 0 & 0 & 0 & | & 0 \end{bmatrix} \begin{bmatrix} 1 & * & * & | & b_1 \\ 0 & 0 & 0 & | & 0 \\ 0 & 0 & 0 & | & 0 \end{bmatrix}$$

In all of the systems considered so far in the MPs, there have been the same number of variables as equations. Equivalently, the coefficient matrix is square. However, it is not always the case that a system of linear equations has the same number of variables as equations, but the analysis of solutions proceeds in the same way.

Example 2.1 *Suppose the following is a reduced form augmented matrix for a system:*

$$\begin{bmatrix} 1 & 0 & 1 & 0 & 0 & | & 5 \\ 0 & 1 & 0 & 0 & -2 & | & 2 \\ 0 & 0 & 0 & 1 & 1 & | & -2 \\ 0 & 0 & 0 & 0 & 0 & | & 0 \end{bmatrix}$$

From the size of the matrix we see that the system has four equations (four rows) and five variables (five columns in the coefficient matrix, none of which is a 0-column). The equations corresponding to this matrix are:

$$x_1 + x_3 = 5 \tag{2.8}$$
$$x_2 - 2x_5 = 2$$
$$x_4 + x_5 = -2$$
$$0 = 0 \tag{2.9}$$

If the bottom equation were $0 = c$ for any $c \neq 0$, there would be no solutions. Since the bottom equation is $0 = 0$, which is always true, it may be ignored. The columns of the matrix without leading entries are #3 and #5, so x_3 and x_5 are the free variables.

Example 2.2 *It is important to be aware that when there are more equations than variables, there may be a row of 0s but still a unique solution (i.e. no free variables). For example, consider the following augmented matrix in reduced form:*

$$\mathbf{R} = \begin{bmatrix} 1 & 0 & 0 & | & 5 \\ 0 & 1 & 0 & | & 2 \\ 0 & 0 & 1 & | & -2 \\ 0 & 0 & 0 & | & 0 \end{bmatrix}$$

The corresponding equations in this case are $x = 5$, $y = 2$, $z = -2$, and $0 = 0$. The last equation is always true and may be ignored, while the other equations specify unique values for x, y, and z. The point here is that in the original system there was a "redundant" equation, which "disappeared" during the row reduction, and really only three equations were necessary to determine the solution.

The only reliable way to determine if a system has infinitely many solutions is to first check whether it has solutions at all (i.e. the reduced system doesn't contain equations of the form $0 = m$ when $m \neq 0$) and then check whether there are columns with no leading variable in the reduced coefficient matrix. The next MP mixes all possible outcomes, and includes non-square matrices.

MP 29 *SOLVING SYSTEMS ALL*

2.3   Applications of Linear Systems

Many real-world applications involve systems of linear equations. In this section we'll see a few of these among the vast number that exist in modern science, economics, and the social sciences. The goal of this section is to develop your instincts for when linear systems are involved in a particular application.

Therefore the focus of this section is on setting up and working with linear systems, not actually finding their solutions, which would involve sending you to a software package such as Maple or Matlab that will numerically solve the systems that you derive. This is the strategy that you will use some day as a STEM professional, but the ability to enter data into a particular software package is not a learning outcome of this text.

Example 2.3 *One application of linear systems involves migration of populations, whether the population is consists of humans, animals, or invasive plants. One example of interest to social scientists and economists is migration of people between cities and the surrounding suburbs or rural areas. This problem is simplified here in order to be addressed by the elementary linear algebra that you have learned. In reality there are many factors that influence shifts in populations, including birth and death rates. This simple model assumes that people are simply moving back and forth between cities and rural areas, and that the rates of movement do not change with time (which is a big and generally unrealistic assumption!). Suppose we start with a city population of c_0 and a rural population of r_0 in the year 2000. We will assume that 7% of the city population moves to rural areas, and 4% of the rural population moves to cities in each 10 year period. We will let c_n and r_n denote the city and rural populations after $10n$ years. In 2010, we have $c_1 = .93c_0 + .04r_0$ and $r_1 = .07c_0 + .96r_0$. These, of course, are linear equations in the variables c_0 and r_0. There are many questions of interest to economists and social scientists involving problems of this sort, but the most basic one is to use this simple model to find the values of c_n and r_n for future times*
The above two linear equations above have coefficient matrix

$$\mathbf{P} = \begin{bmatrix} .93 & .04 \\ .07 & .96 \end{bmatrix}$$

This matrix represents the redistribution of the populations after 10 years have passed. Multiplying the matrix once by the initial distribution lets us compute the column vector

$$\begin{bmatrix} c_1 \\ r_1 \end{bmatrix} = \mathbf{P} \begin{bmatrix} c_0 \\ r_0 \end{bmatrix} = \begin{bmatrix} .93 & .04 \\ .07 & .96 \end{bmatrix} \begin{bmatrix} c_0 \\ r_0 \end{bmatrix} = \begin{bmatrix} .93c_0 + .04r_0 \\ .07c_0 + .96r_0 \end{bmatrix}$$

If we want to know the distribution of these populations in 2020, we multiply by $\mathbf{P}$ again:

$$\begin{bmatrix} c_2 \\ r_2 \end{bmatrix} = \begin{bmatrix} .93 & .04 \\ .07 & .96 \end{bmatrix} \begin{bmatrix} c_1 \\ r_1 \end{bmatrix}$$

$$= \begin{bmatrix} .93 & .04 \\ .07 & .96 \end{bmatrix} \begin{bmatrix} .93 & .04 \\ .07 & .96 \end{bmatrix} \begin{bmatrix} c_0 \\ r_0 \end{bmatrix} = \mathbf{P}^2 \begin{bmatrix} c_0 \\ r_0 \end{bmatrix}$$

As I'm sure you will believe, but which can be proved using a technique called "mathematical induction"that we will discuss later, $10n$ years after 2000, the *distribution of the populations will be*

$$\begin{bmatrix} c_n \\ r_n \end{bmatrix} = \mathbf{P}^n \begin{bmatrix} c_0 \\ r_0 \end{bmatrix}$$

An important question now becomes: after a long period of time, does this distribution "stabilize" in the sense that the populations approach fixed percentages. For example, under some circumstances one might learn that the percentage of people living rural areas tends to 0 after a long period of time. If this is considered an undesirable outcome, one could enact policies intended to change the coefficients and therefore the trend. Even if the rural percentage does not tend to 0, knowing the longer-term trend can help with planning for suitable infrastructure, among other things. This question is precisely equivalent to whether the matrix $\mathbf{P}^n$ "stabilizes" in the sense of approaching some fixed matrix as n gets large. "Approaching", in turn is made precise by the notion of limits from calculus. In this case it is possible to check that the limit of the entries of $\mathbf{P}^n$ is the following matrix (rounded) as $n \to \infty$:

$$\begin{bmatrix} 0.36 & 0.36 \\ 0.64 & 0.64 \end{bmatrix}$$

This means that the population vector stabilizes at approximately

$$\begin{bmatrix} 0.36 & 0.36 \\ 0.64 & 0.64 \end{bmatrix} \begin{bmatrix} c_0 \\ r_0 \end{bmatrix}$$

Definition 2.2 *The matrix $\mathbf{P}$ is an example of a stochastic matrix, which by definition means that the entries of any column add up to 1.*

Stochastic matrices are ubiquitous in science and economics because the columns represent the distribution of certain quantities, which must add up to 1.

Exercise 2.2 *Create your own version of Example 2.3, related to something of interest to you. That is, describe the situation and create a stochastic matrix that represents the incremental change in distribution. You should write out a full explanation in your own words explaining how to compute the population after n time increments.*

Example 2.4 *Spread of contagious diseases can also be modeled using linear algebra. As you may recall, during the COVID pandemic, many governments instituted contact tracing. Suppose, for example, that two people have contracted COVID. Another group of five uninfected people has been in possible contact with those two. Using contact information we can construct a 2×5 contact matrix like the following:*

$$\mathbf{C} = \begin{bmatrix} 1 & 1 & 0 & 0 & 1 \\ 0 & 1 & 1 & 1 & 0 \end{bmatrix}$$

Here each row corresponds to one of the two infected people and each column corresponds to one of the five uninfected people who may have been in contact. There is a 1 in entry c_{ij} if the ith infected person has come into contact with the jth uninfected person. Now suppose that a third group of four uninfected people may have been in contact with the second group of five and we similarly construct a contact matrix

$$\mathbf{D} = \begin{bmatrix} 1 & 1 & 0 & 1 \\ 0 & 1 & 0 & 1 \\ 0 & 0 & 1 & 0 \\ 0 & 0 & 1 & 1 \\ 0 & 1 & 1 & 0 \end{bmatrix}$$

Suppose we are interested in "second order" contact—that is, the number of contacts each person in the last group had with someone in the second group who had contact with someone in the first group. This second order contact is obtained by the matrix product

$$\mathbf{CD} = \begin{bmatrix} 1 & 3 & 1 & 2 \\ 0 & 1 & 2 & 2 \end{bmatrix}$$

Each integer in this matrix represents a second order "path" from one of the original two infected people to each of the four people in the third group. As you can imagine, actual contact matrices can be enormous, which is why it is important to have efficient methods to do matrix calculations on extremely large matrices.

Exercise 2.3 *Give a variation on the above example. That is, create a contact matrix from one group of people to another, with contact from the second group to a third, and compute the matrix of secondary contacts. Your groups should have 2-4 people, and you should explain, in your own words, what the matrices represent.*

Example 2.5 *An important application of linear algebra is network analysis— used in communications, electrical engineering, and other fields. Systems of linear equations occur naturally in networks that are subject to a basic con- servation law: that the sum of inputs to a node is equal to the sum of outputs. A standard example is Kirchhoff's law of currents: that in any electrical net- work, the total current entering a node is equal to the total current leaving it. Many linear algebra texts try to mix it up a bit by including resistors in electrical networks, in which case one has to compute currents using Ohm's law $V = IR$ in order to set up linear equations involving current. This text is intended for students with many interests, and being able to do the necessary 5th grade arithmetic to compute I from R and V is not a learning outcome of this text. Moreover, the electrical engineering students among you will no doubt see even more complex versions of this problem in the future. At any rate, consider the network in Figure 2.1, in which the arrows show input and output to each node, which we assume is subject to a conservation law*

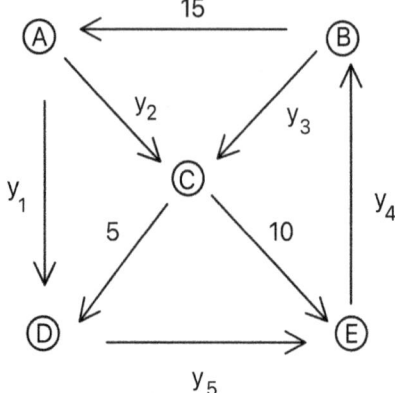

FIGURE 2.1
Network

Each node (labeled A-E) results in a linear equation by setting the total input equal to the total output. We have the following linear equations corresponding to each node:

$$
\begin{array}{ll}
A & 15 = y_1 + y_2 \\
B & y_4 = y_3 + 15 \\
C & y_2 + y_3 = 15 \\
D & y_1 + 5 = y_5 \\
E & y_5 + 10 = y_4
\end{array}
$$

Rearranging we have the following system:

$$
\begin{aligned}
y_1 + y_2 &= 15 \\
y_3 - y_4 &= -15 \\
y_2 + y_3 &= 15 \\
y_1 - y_5 &= -5 \\
y_4 - y_5 &= 10
\end{aligned}
$$

The resulting matrix is

$$
\left[
\begin{array}{ccccc|c}
1 & 1 & 0 & 0 & 0 & 15 \\
0 & 0 & 1 & -1 & 0 & -15 \\
0 & 1 & 1 & 0 & 0 & 15 \\
1 & 0 & 0 & 0 & -1 & -5 \\
0 & 0 & 0 & 1 & -1 & 10
\end{array}
\right]
$$

The reduced form of the matrix is

$$
\begin{bmatrix}
1 & 1 & 0 & 0 & 0 & | & 15 \\
0 & 1 & 1 & 0 & 0 & | & 15 \\
0 & 0 & 1 & -1 & 0 & | & -15 \\
0 & 0 & 0 & 1 & -1 & | & 10 \\
0 & 0 & 0 & 0 & 0 & | & 0
\end{bmatrix}
$$

This system has infinitely many solutions—as is completely reasonable in realistic situations! For example, if this network represents an electrical circuit, each of the currents y_i would depend on the current being applied to the circuit.

Example 2.6 *Now suppose that we tweak the network, replacing the 10 between nodes C and E by a 5. This looks like a perfectly reasonable network, but going through the same steps results in the reduced augmented matrix*

$$
\begin{bmatrix}
1 & 1 & 0 & 0 & 0 & | & 15 \\
0 & 1 & 1 & 0 & 0 & | & 15 \\
0 & 0 & 1 & -1 & 0 & | & -15 \\
0 & 0 & 0 & 1 & -1 & | & 10 \\
0 & 0 & 0 & 0 & 0 & | & -5
\end{bmatrix}
$$

This new system has no solutions! So if your idea to solve a problem in electrical engineering included this network as part of your circuit, you would have proof that your idea would not work—and you would need to try something else. Ruling out unworkable ideas is an important part of science and engineering, and mathematical theorems can definitely help with that.

Exercise 2.4 *In Example 2.5, identify the only column without a leading entry, set the corresponding variable equal to 3, and compute the value of the current y_3. The math is easy—if you load this matrix into some software to answer this, you are wasting a lot of time! (And missing the point.)*

Exercise 2.5 *Construct your own network (different from the one in Example 2.5) with five nodes, ensuring that every node has some input and output. Label some inputs with variables $y_1, y_2, \ldots$ and some with numerical values. Assuming that the network satisfies a conservation law, write down the corresponding system of linear equations. You do not need to write down the matrix or check whether the system actually has solutions.*

2.4 Matrix Form for Systems of Linear Equations

It is very important to have general notation for linear systems and their matrix counterparts, so that we can formally state theorems rather than just considering special cases. We'll start by writing out formally what we have done in the last section. As we saw previously, one way to solve the system is to create the *augmented matrix*

$$\left[\begin{array}{cccc|c} a_{11} & a_{12} & \cdots & a_{1n} & b_1 \\ a_{21} & a_{22} & \cdots & a_{2n} & b_2 \\ \vdots & \vdots & \vdots & \vdots & \vdots \\ a_{n1} & a_{n2} & \cdots & a_{nn} & b_n \end{array}\right]$$

for the system and carry out Gauss-Jordan elimination until the coefficient matrix is in reduced echelon form, then read off the solution(s) (or lack thereof) from the resulting simpler system. There is yet another way to view this system if we let $\mathbf{X} = \begin{bmatrix} x_1 \\ x_2 \\ \vdots \\ x_n \end{bmatrix}$ be the column vector containing the variables,

and $\mathbf{B} = \begin{bmatrix} b_1 \\ b_2 \\ \vdots \\ b_n \end{bmatrix}$ be the matrix of the constants on the right side. If we write

out the product $\mathbf{AX}$ we get

$$\begin{bmatrix} a_{11} & a_{12} & \cdots & a_{1n} \\ a_{21} & a_{22} & \cdots & a_{2n} \\ \vdots & \vdots & \vdots & \vdots \\ a_{n1} & a_{n2} & \cdots & a_{nn} \end{bmatrix} \begin{bmatrix} x_1 \\ x_2 \\ \vdots \\ x_n \end{bmatrix} = \begin{bmatrix} a_{11}x_1 + a_{12}x_2 + \cdots + a_{1n}x_n \\ a_{21}x_1 + a_{22}x_2 + \cdots + a_{2n}x_n \\ \vdots \\ a_{n1}x_1 + a_{n2}x_2 + \cdots + a_{nn}x_n \end{bmatrix} \quad (2.10)$$

which is a column vector. Then the system of equations is equivalent to the following equation of column vectors:

$$\begin{bmatrix} a_{11}x_1 + a_{12}x_2 + \cdots + a_{1n}x_n \\ a_{21}x_1 + a_{22}x_2 + \cdots + a_{2n}x_n \\ \vdots \\ a_{n1}x_1 + a_{n2}x_2 + \cdots + a_{nn}x_n \end{bmatrix} = \begin{bmatrix} b_1 \\ b_2 \\ \vdots \\ b_n \end{bmatrix} \quad (2.11)$$

Putting all this together we see that the system can be equivalently written in the following matrix form:

$$\mathbf{AX} = \mathbf{B} \quad (2.12)$$

Long ago you learned to solve linear equations of the form $ax = b$. If $a \neq 0$ then you may divide each side by a, which is the equivalent of multiplying each side by a^{-1}. You probably didn't consider the case when $a = 0$ because then the equation simply reduces to $0 = b$, but it is useful to say more about this case because it can help understand analogous equation (2.12) in matrix algebra. If $b \neq 0$ then the equation $0x = b$ has *no solutions* because $0x = 0$ for any x. If $a = 0$ and $b = 0$ then the equation is $0x = 0$. Every possible x is a solution to this equation, and in particular there are infinitely many solutions.

While it would be tempting to "solve" equation (2.12) by simply multiplying each side by $\mathbf{A}^{-1}$, you already know that $\mathbf{A}^{-1}$ may not exist, even if $\mathbf{A}$ is a square matrix. But if $\mathbf{A}$ is an invertible matrix, then the theorems of matrix algebra tell us that we can solve this equation as $\mathbf{X} = \mathbf{A}^{-1}\mathbf{B}$ (which, as a reminder, may not be the same as $\mathbf{B}\mathbf{A}^{-1}$). There is another important consequence of this calculation. If $\mathbf{A}$ is invertible, then there is only *one possible solution* to this equation (and the system of linear equations that it represents), namely $\mathbf{A}^{-1}\mathbf{B}$.

Theorem 2.4 *If $\mathbf{A}$ is an invertible matrix, then every equation $\mathbf{A}\mathbf{X} = \mathbf{B}$ has a unique solution, namely $\mathbf{X} = \mathbf{A}^{-1}\mathbf{B}$. In particular, the corresponding system of linear equations has a unique solution.*

2.5 Homogeneous Systems

There is a very important special case for matrix equations, namely equations of the form $\mathbf{A}\mathbf{X} = \mathbf{0}$, i.e. when $\mathbf{B}$ is the column vector with all entries equal to $\mathbf{0}$. This matrix equation, and the corresponding system of linear equations, are referred to as *homogeneous*. The first thing to notice about homogeneous equations is that the zero vector $\mathbf{X} = \mathbf{0}$ is always a solution, referred to as the *trivial solution*.

Now suppose we have an equation $\mathbf{A}\mathbf{X} = \mathbf{B}$ and the column vector $\mathbf{P}_0$ is a solution to the equation. This means that $\mathbf{A}\mathbf{P}_0 = \mathbf{B}$. Suppose now that $\mathbf{Z}$ is any solution to the *homogeneous* equation $\mathbf{A}\mathbf{X} = \mathbf{0}$. This means that $\mathbf{A}\mathbf{Z} = \mathbf{0}$. By the algebra rules for matrix multiplication,

$$\mathbf{A}(\mathbf{P}_0 + \mathbf{Z}) = \mathbf{A}\mathbf{P}_0 + \mathbf{A}\mathbf{Z} = \mathbf{B} + \mathbf{0} = \mathbf{B}.$$

In other words, $\mathbf{P}_0 + \mathbf{Z}$ is a solution to the equation $\mathbf{A}\mathbf{X} = \mathbf{B}$. When $\mathbf{B} \neq \mathbf{0}$, this equation, and the corresponding linear system, are called *inhomogeneous*. We have just seen that if we add any *particular solution* $\mathbf{P}_0$ of an inhomogeneous system $\mathbf{A}\mathbf{X} = \mathbf{B}$ to any solution of the *homogeneous* equation $\mathbf{A}\mathbf{X} = \mathbf{0}$ then we always get another solution to the inhomogeneous system $\mathbf{A}\mathbf{X} = \mathbf{B}$.

Next, suppose that $\mathbf{Q}$ is another solution to the inhomogeneneous system $\mathbf{A}\mathbf{X} = \mathbf{B}$. Then

$$\mathbf{A}(\mathbf{P}_0 - \mathbf{Q}) = \mathbf{A}\mathbf{P}_0 - \mathbf{A}\mathbf{Q} = \mathbf{B} - \mathbf{B} = \mathbf{0}.$$

That is, the difference between any two solutions to the inhomogeneous system $\mathbf{A}\mathbf{X} = \mathbf{B}$ is a solution to the homogeneous system $\mathbf{A}\mathbf{X} = \mathbf{0}$. Now we may always write $\mathbf{Q} = \mathbf{P}_0 + (\mathbf{P}_0 - \mathbf{Q})$ and by what we calculated above, $\mathbf{P}_0 - \mathbf{Q}$ is a solution to the homogeneous system $\mathbf{AX} = \mathbf{0}$. Putting all of this together we have proved the following theorem:

Theorem 2.5 *For any inhomogeneous matrix equation $\mathbf{AX} = \mathbf{B}$ with any particular solution $\mathbf{P}_0$, every solution to $\mathbf{AX} = \mathbf{B}$ is of the form $\mathbf{P}_0 + \mathbf{Z}$, where $\mathbf{Z}$ is a solution to the homogeneous system $\mathbf{AX} = \mathbf{0}$.*

The above theorem shows the importance of homogeneous systems of equations. If one knows the entire solution set to the homogeneous system $\mathbf{AX} = \mathbf{0}$, to find all possible solutions to the *inhomogeneous* system $\mathbf{AX} = \mathbf{B}$, one need only find a single particular solution $\mathbf{P}_0$ to $\mathbf{AX} = \mathbf{B}$, and then take its sum with all solutions to the homogeneous system. If $\mathbf{AX} = \mathbf{0}$ has a unique solution, which of course must be $\mathbf{X} = \mathbf{0}$, and $\mathbf{AX} = \mathbf{B}$ has a solution $\mathbf{P}_0$, then by Theorem 2.5, $\mathbf{P}_0$ must be the unique solution to $\mathbf{AX} = \mathbf{B}$. We state this observation as a theorem:

Theorem 2.6 *If a linear system $\mathbf{AX} = \mathbf{B}$ is consistent, then it has the same number of solutions (unique or infinitely many) as the homogeneous system $\mathbf{AX} = \mathbf{0}$.*

We are now in a position to prove the uniqueness theorem promised earlier. To facilitate the proof, here is the general depiction of the augmented matrix of a homogeneous system in reduced form. As explained when this depiction was first introduced in (2.4), some of the 0-rows and free columns may not exist in a given situation. The free columns include the 0-columns on the left and columns containing $*$s. You have seen all of these outcomes in various MPs.

$$
\left[
\begin{array}{ccccccccccccccc|c}
0 & \cdots & 0 & 1 & * & \cdots & * & 0 & * & \cdots & * & \cdots & 0 & * & \cdots & 0 \\
0 & \cdots & 0 & 0 & 0 & \cdots & 0 & 1 & * & \cdots & * & \cdots & 0 & * & \cdots & 0 \\
0 & \cdots & 0 & 0 & 0 & \cdots & 0 & 0 & 0 & \cdots & 0 & \cdots & 1 & * & \cdots & 0 \\
\vdots & \vdots & \vdots & \vdots & \vdots & \vdots & \vdots & \vdots & \vdots & \vdots & \vdots & \vdots & \vdots & \vdots & \vdots & \vdots \\
0 & \cdots & 0 & 0 & 0 & \cdots & 0 & 0 & 0 & \cdots & 0 & \cdots & 0 & 0 & \cdots & 0 \\
\vdots & \vdots & \vdots & \vdots & \vdots & \vdots & \vdots & \vdots & \vdots & \vdots & \vdots & \vdots & \vdots & \vdots & \vdots & \vdots \\
0 & \cdots & 0 & 0 & 0 & \cdots & 0 & 0 & 0 & \cdots & 0 & \cdots & 0 & 0 & \cdots & 0 \\
\end{array}
\right] \tag{2.13}
$$

The proof of this theorem may seem challenging at first, and it is certainly the most complex proof so far in *Interactive Linear Algebra*. If you don't have much experience with proofs, do not expect to understand every word on the first reading. Continue rereading it very slowly, looking at every symbol, until you could explain it to someone else. Understanding the proof will deepen your understanding of reduced echelon form, and will prepare you for logical arguments that you may encounter in the future in mathematics, computer science, and more advanced engineering courses. Throughout the proof you should refer to (2.13).

Theorem 2.7 *The reduced echelon form of any matrix is unique—that is, it does not matter which row operations are used to obtain it.*

Proof. The question at hand is this: Suppose that we have used row operations in two different ways to reduce a matrix $\mathbf{A}$ to obtain matrices $\mathbf{R}$ and $\mathbf{T}$ in reduced echelon form. Can $\mathbf{R}$ and $\mathbf{T}$ be different? This theorem says that the answer is "no"; that is, we will show that $\mathbf{R} = \mathbf{T}$. We will start with a special case: neither $\mathbf{R}$ nor $\mathbf{T}$ has any free columns. This means that the 0-columns in the above depiction do not exist (in either $\mathbf{R}$ or $\mathbf{T}$) and therefore both have a 1 in the upper left corner. That is, $r_{11} = t_{11} = 1$, and all other entries in the first column must be 0. Likewise, since there are no free columns, the free columns between the first 1 and the second 1 from the left do not exist in $\mathbf{R}$ or $\mathbf{T}$, and therefore $r_{22} = t_{22} = 1$, and all other entries in the second column are 0s. In fact, since no columns with $*$ in them exist, both $\mathbf{R}$ and $\mathbf{T}$ look like this:

$$\begin{bmatrix} 1 & 0 & 0 & \cdots & 0 & \cdots & | & 0 \\ 0 & 1 & 0 & \cdots & 0 & \cdots & | & 0 \\ 0 & 0 & 1 & \cdots & 0 & \cdots & | & 0 \\ \vdots & \vdots & \vdots & \vdots & \vdots & \vdots & | & \vdots \\ 0 & 0 & 0 & \cdots & 0 & 1 & | & 0 \\ 0 & 0 & 0 & \vdots & 0 & 0 & | & 0 \\ \vdots & \vdots & \vdots & \vdots & \vdots & \vdots & | & \vdots \\ 0 & 0 & 0 & \vdots & 0 & 0 & | & 0 \end{bmatrix} \tag{2.14}$$

So $\mathbf{R} = \mathbf{T}$.

At this point we have reduced to the case in which at least one of $\mathbf{R}$ or $\mathbf{T}$ has a free column. We will finish the proof *by contradiction*. If you haven't seen a proof by contradiction before, the idea is very simple: assume that the statement that you want to prove is *not true*, and use logic to reach a *contradiction*—that is, a conclusion that you know is false. This contradiction shows that the original statement that you intended to prove is true. For this proof we will assume that $\mathbf{R} \neq \mathbf{T}$ and follow various cases to obtain contradictions. Except for the very last case, the contradiction will always be that $\mathbf{RX} = \mathbf{0}$ and $\mathbf{TX} = \mathbf{0}$ have *different solutions*. This is a contradiction because $\mathbf{R}$ and $\mathbf{T}$ were obtained by applying row operations to $\mathbf{A}$, and by Theorem 2.1, row operations do not change the solutions to a system.

The first case to consider is that there is a leading column $\mathbf{T}^j$, with leading entry $t_{kj} = 1$, while $\mathbf{R}^j$ is a free column. If $(s_1, \ldots, s_n)$ is a solution to $\mathbf{TX} = \mathbf{0}$, then we must have $s_j = \sum_{i \neq j} -t_{ki} s_i$. Now if $(s_1, \ldots, s_n)$ is *not* a solution to $\mathbf{RX} = \mathbf{0}$, we have our contradiction and the proof of this case is finished. If $(s_1, \ldots, s_n)$ *is* a solution to $\mathbf{RX} = \mathbf{0}$ then x_j is a free variable for $\mathbf{RX} = \mathbf{0}$, and we may replace s_j by $s_j + 5$, and there is some solution

$$(s_1', \ldots, s_{j-1}', s_j + 5, s_{j+1} \ldots, s_n) \tag{2.15}$$

for $\mathbf{RX} = \mathbf{0}$. But (2.15) cannot be a solution to $\mathbf{TX} = \mathbf{0}$ because $s_j + 5 \neq \sum_{i \neq j} -t_{ki}s_i$, and again we have a contradiction. (Our choice of 5 here was arbitrary—we just need to add *something* to s_j to make it different.)

We have now reduced to the case in which $\mathbf{R}$ and $\mathbf{T}$ have the exact same free (and therefore leading) columns and free variables $x_{f_1}, \ldots, x_{f_r}$. By the ordering of the leading entries in reduced echelon form, the leading entries must be in the exact same locations in the two matrices. Since $\mathbf{R}$ and $\mathbf{T}$ are different, there must be some i such that the rows $\mathbf{R}_i$ and $\mathbf{T}_i$ are different. In that row, the only possibility for $r_{iq} \neq t_{iq}$ is for $\mathbf{R}^q$ and $\mathbf{T}^q$ to be free columns. That is, for some m, $r_{if_m} \neq t_{if_m}$. If $r_{ij} = t_{ij} = 1$ are the leading entries in $\mathbf{R}_i$ and $\mathbf{T}_i$, then by Theorem 2.3 we can write

$$x_j = \sum_{f_m > j} -r_{if_m} x_{f_m} \quad \text{and} \quad x_j = \sum_{f_m > j} -t_{if_m} x_{f_m}$$

We can find a solution $(s_1, \ldots, s_n)$ for $\mathbf{TX} = \mathbf{0}$ by setting all the free variables equal to 0 except for $x_{f_m} := 1$, and therefore the left equation above gives us $s_j = -t_{if_m}$. If this is not a solution to $\mathbf{RX} = \mathbf{0}$, we have our contradiction. If $(s_1, \ldots, s_n)$ *is* also a solution for $\mathbf{RX} = \mathbf{0}$, then the right equation above gives us $s_j = -r_{if_m}$. This is a contradiction, since $r_{if_m} \neq t_{if_m}$. ∎

A simpler way to express the coefficient matrix in (2.14) is in the form of what is called a *block matrix* $\begin{bmatrix} \mathbf{I} \\ \mathbf{0} \end{bmatrix}$. One obtains a block matrix from a matrix (called the *underlying matrix*) by dividing it into rectangular pieces using horizontal and vertical separations that go fully from side to side and top to bottom. It contains the same information as the underlying matrix, but organizes it differently (and more compactly). Using block notation we can extract what we proved in the first part of the proof of Theorem 2.7:

Theorem 2.8 *A linear system* $\mathbf{AX} = \mathbf{0}$ *has a unique solution if and only if the reduced echelon form of* $\mathbf{A}$ *is of the block form* $\begin{bmatrix} \mathbf{I} \\ \mathbf{0} \end{bmatrix}$.

The same type of caveat applies here: it is possible that the block $\mathbf{0}$ may not not exist, which happens when the matrix $\mathbf{A}$ is square (and the system has a unique solution).

There is an algebraic system involving block matrices. Block matrices can be multiplied if they are of compatible size, and the individual blocks that must be multiplied are also compatible. This is less complicated than it sounds. For example, the product

$$\begin{bmatrix} \mathbf{A} & \mathbf{B} \\ \mathbf{C} & \mathbf{D} \end{bmatrix} \begin{bmatrix} \mathbf{E} & \mathbf{F} & \mathbf{G} \\ \mathbf{H} & \mathbf{J} & \mathbf{K} \end{bmatrix} = \begin{bmatrix} \mathbf{AE} + \mathbf{BH} & \mathbf{AF} + \mathbf{BJ} & \mathbf{AG} + \mathbf{BK} \\ \mathbf{CE} + \mathbf{DH} & \mathbf{CF} + \mathbf{DJ} & \mathbf{CG} + \mathbf{DK} \end{bmatrix}$$

of block matrices will be defined not only because the left block matrix is 2×2 and the right block matrix is 2×3 but also provided that the necessary matrix products on the right, such as $\mathbf{AE}$ and $\mathbf{CH}$, are also defined. So there are "two levels of compatibility" that must be considered. It is also an important fact that the underlying matrix on the right is the product of the underlying matrices on the left. The proof that block matrix multiplication works "as advertised" is quite tedious and doesn't contribute much to understanding; therefore I will omit it. But we will use block matrix multiplication at several important places in this text, and it is used frequently in physics and engineering.

Exercise 2.6 *Divide the matrix*

$$\mathbf{A} = \begin{bmatrix} 1 & 0 & 0 & 1 \\ 0 & 1 & 0 & 1 \\ 0 & 0 & 0 & 1 \end{bmatrix}$$

*into block matrices in the following ways. If it is **not possible** to do so, state this and explain why. For example, here is a division of* $\mathbf{A}$ *into a block matrix that contains* $\mathbf{I}_{2\times2}$, *and you should convince yourself that this is the only possible way to do so.*

$$\begin{bmatrix} \begin{bmatrix} 1 & 0 \\ 0 & 1 \end{bmatrix} & \begin{bmatrix} 0 & 1 \\ 0 & 1 \end{bmatrix} \\ \begin{bmatrix} 0 & 0 \end{bmatrix} & \begin{bmatrix} 0 & 1 \end{bmatrix} \end{bmatrix}$$

1. *A* 2×2 *block matrix that includes* $[0]$ *as a block.*
2. *A* 3×3 *block matrix.*
3. *A block matrix that includes a* 2×3 *block and a* 1×2 *block.*

Exercise 2.7 *Suppose that* $\mathbf{B}$ *is a* 2×3 *matrix,* $\mathbf{E}$ *and* $\mathbf{F}$ *are matrices of the same size, and the underlying matrix of the block matrix* $\begin{bmatrix} \mathbf{E} & \mathbf{F} \\ \mathbf{G} & \mathbf{H} \end{bmatrix}$ *is* 6×8. *In order for the following block matrix product to be defined, what size must* $\mathbf{G}$ *be?*

$$\begin{bmatrix} \mathbf{A} & \mathbf{B} \\ \mathbf{C} & \mathbf{D} \end{bmatrix} \begin{bmatrix} \mathbf{E} & \mathbf{F} \\ \mathbf{G} & \mathbf{H} \end{bmatrix}$$

Exercise 2.8 *Check the block multiplication formula* $\begin{bmatrix} 1 & 0 \\ 0 & \mathbf{R} \end{bmatrix}^{-1} = \begin{bmatrix} 1 & 0 \\ 0 & \mathbf{R}^{-1} \end{bmatrix}$ *where* $\mathbf{R}$ *is an invertible matrix. To do so, note that the identity matrix can be written in block form as* $\begin{bmatrix} 1 & 0 \\ 0 & \mathbf{I} \end{bmatrix}$ *where* $\mathbf{I}$ *is the identity matrix of one size smaller. Your work should include checking the compatibility of any of the blocks that you multiply, including the 0-blocks, which in this case are row and column vectors.*

Remark 2.1 *If you have been paying extremely close attention, you might have noticed that I studiously avoided phrases like "**the** reduced echelon form of* $\mathbf{A}$*" until we proved, in Theorem 2.7, that this form is unique. This uniqueness is essential to make the following definition.*

Definition 2.3 *A linear system* $\mathbf{AX} = \mathbf{B}$ *is called overdetermined if the reduced echelon form* $\mathbf{R}$ *of* $\mathbf{A}$ *has more rows than columns, meaning that there are more equations than variables in the corresponding system. The system is underdetermined if* $\mathbf{R}$ *has more columns than rows, meaning there are more variables than equations.*

In textbooks at this level, these terms are sometimes improperly applied to the original matrix $\mathbf{A}$. The issue with doing so is that there could be redundant equations buried in $\mathbf{AX} = \mathbf{B}$. This is seen in Example 2.2, in which the original matrix $\mathbf{A}$ was a 4×3 matrix. ($\mathbf{A}$ was not given in the example, but whatever it is, it must be a 3×4 and could certainly have all non-zero rows.) In that example, the system corresponding to $\mathbf{R}$ had the corresponding equation $0 = 0$, which can be ignored. In particular, that system was *not* overdetermined despite the fact that the original matrix $\mathbf{A}$ was an 4×3 matrix.

Theorem 2.9 *Let* $\mathbf{AX} = \mathbf{0}$ *be a system of linear equations. If the system is underdetermined then there are infinitely many solutions of* $\mathbf{AX} = \mathbf{0}$. *If the system is overdetermined then there is an inconsistent system* $\mathbf{AX} = \mathbf{W}$ *for some column vector* $\mathbf{W}$.

Proof. If the reduced echelon form $\mathbf{R}$ of $\mathbf{A}$ has more columns than rows, there must be at least one free column, so there are infinitely many solutions. If $\mathbf{R}$ has more rows than columns then the bottom row of $\mathbf{R}$ must be a 0-row (why?). Therefore, if

$$\mathbf{W} = \begin{bmatrix} 0 \\ \vdots \\ 0 \\ b \end{bmatrix}$$

for any $b \neq 0$, then $\mathbf{AX} = \mathbf{W}$ is an inconsistent system. ∎

2.6   Parameterized Solutions

We will now revisit the problem mentioned earlier. If a system has infinitely many solutions, how do we describe them all? One very good way to do so is through *parameterized solutions*. Let's revisit the reduced form outcomes of possible augmented 3×4 matrices with no 0-columns (i.e. every column corresponds to a variable):

$$\begin{bmatrix} 1 & 0 & a & | & b_1 \\ 0 & 1 & b & | & b_2 \\ 0 & 0 & 0 & | & 0 \end{bmatrix} \quad \begin{bmatrix} 1 & a & 0 & | & b_1 \\ 0 & 0 & 1 & | & b_2 \\ 0 & 0 & 0 & | & 0 \end{bmatrix} \quad \begin{bmatrix} 1 & a & b & | & b_1 \\ 0 & 0 & 0 & | & 0 \\ 0 & 0 & 0 & | & 0 \end{bmatrix}$$

In the left matrix, the third column has no leading entry, and if our variables are x, y, z then z is the one and only free variable. It is useful to rename z as a reminder that z is a free variable. For this purpose we will set $z = t$, and we will call t the *parameter* for z. This leaves us with three *parameterized equations*: $x + at = b_1$ and $y + bt = b_2$ and $z = t$. We may now solve for the remaining two variables: $y = b_2 - bt$ and $x = b_1 - at$. There are infinitely many solutions because we may choose any value for t, which then determines values for x and y. For example, if we choose $t = 0$ then we have a *particular solution* $(b_1, b_2, 0)$. If we choose $t = -1$ then we have a particular solution $(b_1 + a, b_2 + b, -1)$, and so on.

Exercise 2.9 *Carry out the same analysis for the middle matrix above: find the free variable, set it equal to t, solve for the remaining two variables, and then find a particular solution for one value of t other than $0, 1, -1$.*

For the last matrix, there are two free columns, corresponding to variables y and z. Now we'll need two parameters, and we'll choose $y = r$ and $z = s$. This gives us one equation $x + ar + bs = b_1$, or $x = b_1 - ar - bs$, along with $y = r$ and $z = s$. To find a particular solution we take particular values of r and s—for example, $r = 2$ and $s = 3$ gives us a solution $(b_1 - 2a - 3b, 2, 3)$.

It is important to remember that the replacement of variables by parameters serves only the purpose of helping humans keep everything straight—it has no mathematical meaning. Mathematically speaking, it doesn't matter what letters you choose, and the letters we are choosing are simply related to the traditional choices that scientists and engineers make. If you are writing software, the computer doesn't care about things like that, and it might be simpler just to not to rename the variables at all!

We have essentially been carrying out the details of Theorem 2.5 using these parameterizations. To see this in action, let's revisit the reduced augmented matrix from Example 2.1:

$$\left[\begin{array}{ccccc|c} 1 & 0 & 1 & 0 & 0 & 5 \\ 0 & 1 & 0 & 0 & -2 & 2 \\ 0 & 0 & 0 & 1 & 1 & -2 \\ 0 & 0 & 0 & 0 & 0 & 0 \end{array}\right]$$

This matrix corresponds to the equation $\mathbf{AX} = \mathbf{B}$, where

$$A = \begin{bmatrix} 1 & 0 & 1 & 0 & 0 \\ 0 & 1 & 0 & 0 & -2 \\ 0 & 0 & 0 & 1 & 1 \\ 0 & 0 & 0 & 0 & 0 \end{bmatrix} \quad X = \begin{bmatrix} x_1 \\ x_2 \\ x_3 \\ x_4 \\ x_5 \end{bmatrix} \quad B = \begin{bmatrix} 5 \\ 2 \\ -2 \\ 0 \end{bmatrix}$$

Since the third and fifth columns have no leading entries, we parameterize $x_3 = r$ and $x_5 = s$. We have the following additional parameterized equations:

$$x_1 + r = 5 \tag{2.16}$$
$$x_2 - 2s = 2$$
$$x_4 + s = -2$$

To get a *particular solution,* you are "free" to choose any values for r and s. For example, $r = 2$ and $s = -1$ giving a particular solution

$$x_1 = 3, x_2 = 0, x_3 = 2, x_4 = -1, x_5 = -1$$

The final step is to rewrite all of this back in matrix form. To do so, solve for each of the variables in terms of the parameters using the equations (2.16) and write it as a column matrix to obtain a *parametrized solution*:

$$x_1 = 5 - r$$
$$x_2 = 2 + 2s$$
$$x_3 = r$$
$$x_4 = -2 - s$$
$$x_5 = s$$

In matrix form this equation is

$$\mathbf{X} = \begin{bmatrix} 5 - r \\ 2 + 2s \\ r \\ -2 - s \\ s \end{bmatrix}$$

We now rewrite $\mathbf{X}$, explicitly including all coefficients:

$$\mathbf{X} = \begin{bmatrix} 5 + (-1)r + 0s \\ 2 + 0r + 2s \\ 0 + 1r + 0s \\ -2 + 0r + (-1)s \\ 0 + 0r + 1s \end{bmatrix}$$

This can be distributed as a sum

$$\mathbf{X} = \begin{bmatrix} 5 \\ 2 \\ 0 \\ -2 \\ 0 \end{bmatrix} + r \begin{bmatrix} -1 \\ 0 \\ 1 \\ 0 \\ 0 \end{bmatrix} + s \begin{bmatrix} 0 \\ 2 \\ 0 \\ -1 \\ 1 \end{bmatrix} \tag{2.17}$$

This is the matrix form of the parameterized solution to $\mathbf{AX} = \mathbf{B}$. Often the most convenient particular solution comes from taking $r = s = 0$. In this case we obtain

$$\mathbf{P}_0 = \begin{bmatrix} 5 \\ 2 \\ 0 \\ -2 \\ 0 \end{bmatrix}$$

which is the first (constant) term of Equation (2.17). So now we have

$$\mathbf{X} = \mathbf{P}_0 + r \begin{bmatrix} -1 \\ 0 \\ 1 \\ 0 \\ 0 \end{bmatrix} + s \begin{bmatrix} 0 \\ 2 \\ 0 \\ -1 \\ 1 \end{bmatrix}$$

The sum of the right two summands of this equation consist of all possible solutions to the homogeneous system $\mathbf{AX} = \mathbf{0}$ and therefore we have written all solutions $\mathbf{X}$ as $\mathbf{P}_0 + \mathbf{Z}$, where $\mathbf{Z}$ is any possible solution to the homogeneous system $\mathbf{AX} = \mathbf{0}$. That is, we have explicitly carried out what Theorem 2.5 tells us must happen.

Exercise 2.10 *Explain the statement, "The sum of the right two summands of this equation consist of all possible solutions to the homogeneous system* $\mathbf{AX} = \mathbf{0}$*." That is, why is that sum a solution to the homogeneous system, and why (for all choices of r and s) does it constitute all of them?*

MP 30 *SOLVING SYSTEMS II*

We conclude this section by adding some terminology that you might see elsewhere–much of it rooted in strategies for working things out "by hand". First, "echelon form" is the same as "reduced echelon form" except that the leading entries need not be 1 (some people require this, however) and the entries in the column above a leading entry need not be zero. From a computational standpoint (especially "by hand"), echelon form mainly serves as an intermediate form, after which the reduction to reduced echelon form is called "backsolving". Leading entries may also be referred to as "pivots". Here is a matrix in echelon form:

$$\begin{bmatrix} 5 & 3 & 1 & 4 & 1 \\ 0 & 0 & 2 & 1 & -1 \\ 0 & 0 & 0 & 0 & 4 \end{bmatrix}$$

The "backsolving" part is what you have done in MPs to get 0s above the leading entries to achieve reduced echelon form, and if necessary multiplying rows by scalars to make the leading entries 1.

2.7   Elementary Matrices

Definition 2.4 *If R is any row operation, the elementary matrix* $\mathbf{E}_R$ *is defined to be the matrix obtained by applying the R to the identity matrix* $\mathbf{I}$ *(of any size).* $\mathbf{E}_R$ *is called the elementary matrix corresponding to the row operation R.*

The subscript "R" in $\mathbf{E}_R$ is used when it is important to know exactly which row operation we are talking about, but if the particular row operation isn't specified, we will often simply denote elementary matrices without any subscript R. This is especially useful when we are discussing several elementary matrices, which we will denote by $\mathbf{E}_1, \ldots, \mathbf{E}_n$ rather than $\mathbf{E}_{R_1}, \ldots, \mathbf{E}_{R_n}$.

The identity matrix $\mathbf{I}$ is itself an elementary matrix, since it may be obtained from itself using the row operation that multiplies any row by 1. If R is the row operation that swaps the first and third rows, then the corresponding 3×3 and 4×4 elementary matrices are

$$\begin{bmatrix} 0 & 0 & 1 \\ 0 & 1 & 0 \\ 1 & 0 & 0 \end{bmatrix} \text{ and } \begin{bmatrix} 0 & 0 & 1 & 0 \\ 0 & 1 & 0 & 0 \\ 1 & 0 & 0 & 0 \\ 0 & 0 & 0 & 1 \end{bmatrix}$$

Likewise the elementary matrices of these sizes corresponding to the row operation that multiplies the second row by 3 are

$$\begin{bmatrix} 1 & 0 & 0 \\ 0 & 3 & 0 \\ 0 & 0 & 1 \end{bmatrix} \text{ and } \begin{bmatrix} 1 & 0 & 0 & 0 \\ 0 & 3 & 0 & 0 \\ 0 & 0 & 1 & 0 \\ 0 & 0 & 0 & 1 \end{bmatrix}$$

And the elementary matrices of these sizes corresponding to adding 5 times the second row to the top row are

$$\begin{bmatrix} 1 & 5 & 0 \\ 0 & 1 & 0 \\ 0 & 0 & 1 \end{bmatrix} \text{ and } \begin{bmatrix} 1 & 5 & 0 & 0 \\ 0 & 1 & 0 & 0 \\ 0 & 0 & 1 & 0 \\ 0 & 0 & 0 & 1 \end{bmatrix}$$

It is a remarkable fact every elementary row operation R may be carried out by multiplication on the left by the elementary matrix $\mathbf{E}_R$. Put another way, elementary matrices follow a variant of the golden rule: they "do unto other matrices as was done unto them". The next theorem formally states this fact.

Theorem 2.10 *If* $\mathbf{E}_R$ *is the is the elementary matrix corresponding to the row operation R, and* $\mathbf{E}_R \mathbf{A}$ *is defined, then* $\mathbf{E}_R \mathbf{A}$ *is the matrix obtained from* $\mathbf{A}$ *by applying the row operation R.*

Let's test this for the row operation consisting of multiplying the third row of a matrix by the scalar 3, applied to a 3×3 matrix. If we perform this operation on the identity matrix, we get the elementary matrix $\mathbf{E} :=$ $\begin{bmatrix} 1 & 0 & 0 \\ 0 & 3 & 0 \\ 0 & 0 & 1 \end{bmatrix}$. To try out theorem, we now multiply $\mathbf{E}$ (on the left) by another

3×3 matrix $\mathbf{A} = \begin{bmatrix} a & b & c \\ d & e & f \\ g & h & k \end{bmatrix}$:

$$\begin{bmatrix} 1 & 0 & 0 \\ 0 & 3 & 0 \\ 0 & 0 & 1 \end{bmatrix} \begin{bmatrix} a & b & c \\ d & e & f \\ g & h & k \end{bmatrix} = \begin{bmatrix} a & b & c \\ 3d & 3e & 3f \\ g & h & k \end{bmatrix}$$

As you can see, the effect on $\mathbf{A}$ of left multiplication by $\mathbf{E}$ is precisely the row operation of multiplying the middle row of $\mathbf{A}$ by 3.

Likewise, adding twice the top row of $\mathbf{I}$ to the bottom row results in a new matrix $\begin{bmatrix} 1 & 0 & 0 \\ 0 & 1 & 0 \\ 2 & 0 & 1 \end{bmatrix}$. This time we will try a 3×2 matrix for $\mathbf{A}$:

$$\begin{bmatrix} 1 & 0 & 0 \\ 0 & 1 & 0 \\ 2 & 0 & 1 \end{bmatrix} \begin{bmatrix} a & b \\ d & e \\ g & h \end{bmatrix} = \begin{bmatrix} a & b \\ d & e \\ 2a + g & 2b + h \end{bmatrix}$$

Again you see that the effect on $\mathbf{A}$ is to add twice its top row to its bottom row—and it should be clear that size doesn't matter as long as $\mathbf{EA}$ is defined. These examples are representative of what happens in general, but let's go ahead and see a more formal proof of Theorem 2.10 for the elementary row operation R that consists of adding the product of a scalar $c \neq 0$ times the ith row to the kth row. The corresponding elementary matrix $\mathbf{E}$ is obtained from $\mathbf{I}$ by performing this row operation. Here is $\mathbf{E}$, where i and k are showing the locations of the ith and k^{th} columns and rows.

$$\mathbf{E} = \begin{array}{c} \\ \\ i \\ \\ k \\ \\ \\ \end{array} \begin{bmatrix} & & i & & k & & \\ 1 & \cdots & 0 & \cdots & 0 & \cdots & 0 \\ \vdots & \vdots & \vdots & \vdots & \vdots & \vdots & \vdots \\ 0 & \cdots & 1 & \cdots & 0 & \cdots & 0 \\ \vdots & \vdots & \vdots & \vdots & \vdots & \vdots & \vdots \\ 0 & \cdots & c & \cdots & 1 & \cdots & 0 \\ \vdots & \vdots & \vdots & \vdots & \vdots & \vdots & \vdots \\ 0 & \cdots & 0 & \cdots & 0 & \cdots & 1 \end{bmatrix} \qquad (2.18)$$

Since only the kth row has changed from the identity matrix, in the product
EA, the only entries that will change are entries

$$(\mathbf{EA})_{kj} = \mathbf{E}_k \cdot \mathbf{A}^j \tag{2.19}$$

$$= (0, \ldots, c, \ldots, 1, \ldots, 0) \cdot (a_{1k}, \ldots, a_{nk}) = ca_{ik} + a_{jk}$$

But the entry $(\mathbf{EA})_{kj}$ is just the jth entry of the kth row of **EA**, and the
above equation therefore precisely says that the kth row of **EA** is obtained
from **A** by the specified row operation R. The rest of the proof is an exercise.

Performing repeated row operations corresponds to repeatedly multiplying
by the corresponding elementary matrices on the left. For example, if the first
row operation applied to **A** is to swap the first and third rows of a $3 \times k$
matrix, followed by adding 3 times the second row to the bottom row, then
one would multiply by the following two matrices on the left, in the order
shown:

$$\begin{bmatrix} 1 & 0 & 0 \\ 0 & 1 & 0 \\ 0 & 3 & 1 \end{bmatrix} \begin{bmatrix} 0 & 0 & 1 \\ 0 & 1 & 0 \\ 1 & 0 & 0 \end{bmatrix} \mathbf{A}$$

The order (right-most matrix corresponds to first row operation) is important
here, because matrix multiplication is generally not commutative, even for
some elementary matrices.

Exercise 2.11 *Show that these two elementary matrices do not commute.
You don't have to do the whole product, just find a particular entry that is
different on the left and the right.*

$$\begin{bmatrix} 1 & 0 & 0 \\ 0 & 1 & 0 \\ 0 & 3 & 1 \end{bmatrix} \begin{bmatrix} 0 & 0 & 1 \\ 0 & 1 & 0 \\ 1 & 0 & 0 \end{bmatrix} \neq \begin{bmatrix} 0 & 0 & 1 \\ 0 & 1 & 0 \\ 1 & 0 & 0 \end{bmatrix} \begin{bmatrix} 1 & 0 & 0 \\ 0 & 1 & 0 \\ 0 & 3 & 1 \end{bmatrix}$$

MP 31 *ELEMENTARY MATRICES*

Theorem 2.11 *Every elementary matrix is invertible and its inverse is an
elementary matrix.*

To prove this theorem, note that every row operation can be "reversed" by
another row operation. For example, to reverse the operation R of multiplication of the ith by a non-zero scalar s, use the row operation R' that multiplies
the same row by $\frac{1}{s}$. Then $\mathbf{E}_{R'}\mathbf{E}_R$ is obtained from **I** by first multiplying the
i^{th} row of **I** by s, and then multiplying the same row by $\frac{1}{s}$, which leaves the
identity. This shows $\mathbf{E}_{R'}\mathbf{E}_R = \mathbf{I}$, and the proof that $\mathbf{E}_R\mathbf{E}_{R'} = \mathbf{I}$ is similar.

Exercise 2.12 *Describe the "inverse operations" to the two remaining row
operations.*

Exercise 2.13 *Show that the elementary matrix for the row operation that adds a scalar multiple of one row to a row **below it** is lower triangular. For this, you need only explain where the c is in the matrix (2.18).*

We will need the next theorem later.

Theorem 2.12 *Let* $\mathbf{E}$ *be an elementary matrix. If when multiplied on the left,*

    1. $\mathbf{E}$ *swaps two rows or multiplies a row by a non-zero scalar, then* $\mathbf{E} = \mathbf{E}^T$.

    2. $\mathbf{E}$ *adds scalar c times row i to row k then* $\mathbf{E}^T$ *is the elementary matrix that adds scalar c times row k to row i.*

In all cases, the transpose of an elementary matrix is an elementary matrix.

Proof. The simplest case is an elementary matrix that multiplies a row of $\mathbf{I}$ by a scalar. The resulting matrix $\mathbf{E}$ is diagonal, hence symmetric, i.e. $\mathbf{E} = \mathbf{E}^T$. For the other two operations, note that the transpose of $\mathbf{E}$ is obtained by applying, to the identity matrix, the operations to columns, rather than rows. We will refer to these as *column operations*. The elementary matrix that swaps rows i and k rows looks like this:

$$\mathbf{E} = \begin{bmatrix} & & i & & k & & \\ 1 & \cdots & 0 & \cdots & 0 & \cdots & 0 \\ \vdots & \vdots & \vdots & \vdots & \vdots & \vdots & \vdots \\ i \quad 0 & \cdots & 0 & \cdots & 1 & \cdots & 0 \\ \vdots & \vdots & \vdots & \vdots & \vdots & \vdots & \vdots \\ k \quad 0 & \cdots & 1 & \cdots & 0 & \cdots & 0 \\ \vdots & \vdots & \vdots & \vdots & \vdots & \vdots & \vdots \\ 0 & \cdots & 0 & \cdots & 0 & \cdots & 1 \end{bmatrix}$$

As you can readily see, $\mathbf{E}$ is also obtained from the identity matrix by swapping columns i and k and therefore $\mathbf{E}^T$ is also obtained from the identity by swapping rows i and k. That is $\mathbf{E}^T = \mathbf{E}$. As for adding a non-zero scalar c times row i to row k, you can see that $\mathbf{E}$ is also obtained by adding c times column k to column i. This proves the second part.

$$\mathbf{E} = \begin{bmatrix} & & i & & k & & \\ 1 & \cdots & 0 & \cdots & 0 & \cdots & 0 \\ \vdots & \vdots & \vdots & \vdots & \vdots & \vdots & \vdots \\ i \quad 0 & \cdots & 1 & \cdots & 0 & \cdots & 0 \\ \vdots & \vdots & \vdots & \vdots & \vdots & \vdots & \vdots \\ k \quad 0 & \cdots & c & \cdots & 1 & \cdots & 0 \\ \vdots & \vdots & \vdots & \vdots & \vdots & \vdots & \vdots \\ 0 & \cdots & 0 & \cdots & 0 & \cdots & 1 \end{bmatrix}$$

∎

Exercise 2.14 *Write down the 4×4 elementary matrix that carries out each of the following row operations, and show that the statements in Theorem 2.12 true in this case:*

    1. Swapping the second and fourth rows.

    2. Adding 4 times the bottom row to the top row.

Exercise 2.15 *Show that the elementary matrix for swapping two rows can be obtained by repeatedly using only the first two row operations RO1 and RO2. To do so, first show this for a 2×2 matrix, starting with*

$$\begin{bmatrix} 1 & 0 \\ 0 & 1 \end{bmatrix} \to \begin{bmatrix} 1 & 0 \\ 1 & 1 \end{bmatrix} \to \begin{bmatrix} 2 & 1 \\ 1 & 1 \end{bmatrix} \to \cdots$$

Now if you look at the proof of Theorem 2.12, you will see that all of the action occurs in just two rows and two columns. So the final step in this exercise is to take your argument for a 2×2 matrix and insert it into more abstract matrices of the type in the proof of Theorem 2.12.

The theorem that you proved in Exercise 2.15 shows that you don't actually need RO3 to do Gauss-Jordan elimination! The reason that it is traditional to include RO3 is that RO3 carries out this operation in a single easy-to-explain step, which is also generally more efficient than using the first two row operations five times to carry it out!

Exercise 2.16 *Finish the proof of Theorem 2.10. You can finesse the proof for swapping rows using Exercise 2.15. That leaves only the proof for the operation of multiplication by a scalar, which is somewhat simpler than the case proved above: adding a scalar multiple of one row to another.*

2.8   Applications of Elementary Matrices

We are now in a position to confirm the validity of our process for determining whether a matrix is invertible, and if it is invertible, finding its inverse. Recall that the process is to create the augmented matrix $\mathbf{A} \mid \mathbf{I}$ and perform row operations on the augmented matrix until it is in reduced form (which by our definition means that the matrix formerly known as "$\mathbf{A}$", on the left, is in reduced echelon form). The first point to be made is that if you multiply an augmented matrix $\mathbf{A} \mid \mathbf{B}$ on the left by an elementary matrix $\mathbf{E}$, the result is the augmented matrix $\mathbf{EA} \mid \mathbf{EB}$. The proof is easier than it may seem. For example, suppose $\mathbf{E}$ is the elementary matrix that multiplies the nth row by a scalar c. Then $\mathbf{E}(\mathbf{A} \mid \mathbf{B})$ multiplies the nth row of $\mathbf{A} \mid \mathbf{B}$ by c. But that's exactly the same as multiplying the nth row of $\mathbf{A}$ and by c, multiplying the nth row of $\mathbf{B}$ by c, and augmenting them. Now that is the same as the augmentation $\mathbf{EA} \mid \mathbf{EB}$.

Now consider the steps to find whether $\mathbf{A}$ is invertible, performing row operations $R_1, \ldots, R_k$ on $\mathbf{A} \mid \mathbf{I}$ (starting with R_1). If $\mathbf{E}_1, \ldots, \mathbf{E}_k$ are the corresponding elementary matrices, then equivalently we are multiplying the augmented matrix on the left by these row operations:

$$\mathbf{A} \mid \mathbf{I} \to (\mathbf{E}_1 \mathbf{A} \mid \mathbf{E}_1 \mathbf{I}) \to (\mathbf{E}_2 \mathbf{E}_1 \mathbf{A} \mid \mathbf{E}_1 \mathbf{E}_2 \mathbf{I}) \to \cdots \qquad (2.20)$$

$$\to (\mathbf{E}_1 \cdots \mathbf{E}_k \mathbf{A} \mid \mathbf{E}_1 \cdots \mathbf{E}_k \mathbf{I}) \qquad (2.21)$$

The process is finished when the matrix $\mathbf{E}_1 \cdots \mathbf{E}_k \mathbf{A}$ on the left is the reduced echelon form of $\mathbf{A}$. If $\mathbf{E}_1 \cdots \mathbf{E}_n \mathbf{A} = \mathbf{I}$ then since elementary matrices are invertible (Theorem 2.11) then we can write $\mathbf{A} = \mathbf{E}_n^{-1} \cdots \mathbf{E}_1^{-1}$. From Theorem 1.2 we know that the product of invertible matrices is invertible, so $\mathbf{A}$ is invertible, and $\mathbf{A}^{-1} = \mathbf{E}_1 \cdots \mathbf{E}_n$, which is equal to the final right matrix in (2.21). We are now ready to prove the next theorem, which summarizes all that we know at this point about non-singular (invertible) matrices.

Before going on, I need to mention a few facts about logic and proofs. The next theorem says that five statements are "equivalent". This means that if you know that any one of them is true for a matrix $\mathbf{A}$ then all of the rest are true for $\mathbf{A}$. Likewise, if any of them is false for $\mathbf{A}$, then all the rest must be false for $\mathbf{A}$. In many situations it is easier to prove (or disprove) one of the statements than the others. For example, maybe you want to know that $\mathbf{AX} = \mathbf{B}$ has a unique solution. The fastest way to do this might be to show that $\mathbf{A}$ is invertible, and once you have proved that, all of the remaining statements of theorem are true, including the one that you need. There are many ways to prove that a list of statements are equivalent, but a very common one is to find some permutation of implications–just making sure that all implications are proved. We will use a very common strategy below: a cyclic argument, showing that the first statement implies the second, which implies the third, and so on, finally proving that the last statement implies the first. Then every statement implies every other statement. For example, although we explicitly proved (2) $\Rightarrow$ (3), we can also derive (3) $\Rightarrow$ (2) because

$$(3) \Rightarrow (4) \Rightarrow (5) \Rightarrow (1) \Rightarrow (2).$$

Theorem 2.13 *The following are equivalent for a square matrix* $\mathbf{A}$:

1. $\mathbf{A}$ *is invertible.*

2. *Any equation* $\mathbf{AX} = \mathbf{B}$ *has a unique solution.*

3. *The homogeneous equation* $\mathbf{AX} = \mathbf{0}$ *has a unique solution.*

4. *The reduced echelon form of* $\mathbf{A}$ *is* $\mathbf{I}$.

5. $\mathbf{A}$ *is a product of elementary matrices.*

Proof. By Theorem 2.4, if $\mathbf{A}$ is invertible then any equation $\mathbf{AX} = \mathbf{B}$ has a unique solution, proving (1) $\Rightarrow$ (2). Since we may take $\mathbf{B} = \mathbf{0}$ in this equation, (2) $\Rightarrow$ (3). Since the system of equations for the reduced form of the augmented matrix $\mathbf{A} \mid \mathbf{0}$ has the same unique solution as the original linear system, the reduced echelon form of $\mathbf{A}$ must be the identity. This shows (3) $\Rightarrow$ (4). If (4) is true then by the argument at the beginning of this section, we know that $\mathbf{A}$ is a product of elementary matrices. Finally, if $\mathbf{A}$ is a product of elementary matrices then since elementary matrices are invertible (Theorem 2.11) and products of invertible matrices are invertible (Theorem 1.2), $\mathbf{A}$ is invertible.
∎

The next theorem was promised long ago:

Theorem 2.14 *If $\mathbf{A}$ is a square matrix and there is a matrix $\mathbf{B}$ such that either $\mathbf{AB} = \mathbf{I}$ or $\mathbf{BA} = \mathbf{I}$, then $\mathbf{A}$ is invertible (and of course $\mathbf{A}^{-1} = \mathbf{B}$).*

Proof. Suppose that $\mathbf{BA} = \mathbf{I}$. By Theorem 2.13, to show $\mathbf{A}$ is invertible we need only show that the equation $\mathbf{AX} = \mathbf{0}$ has a unique solution. Suppose $\mathbf{X}_0$ is a solution. Multiplying both sides on the left by the matrix $\mathbf{B}$ gives us

$$\mathbf{B}(\mathbf{AX}_0) = \mathbf{B0} \Leftrightarrow (\mathbf{BA})\mathbf{X}_0 = \mathbf{0} \Leftrightarrow \mathbf{IX}_0 = \mathbf{0} \Leftrightarrow \mathbf{X}_0 = \mathbf{0}$$

So the unique solution of the system is $\mathbf{0}$. The rest of the proof is an exercise.
∎

Exercise 2.17 *Finish the proof of Theorem 2.14 by doing the case when $\mathbf{AB} = \mathbf{I}$. Hint: You don't have to repeat the argument for the first case, although that would be a useful exercise itself. Instead, apply the case that was done above to the matrix $\mathbf{B}$. Once you know that $\mathbf{B}$ has an inverse, use it!*

Elementary matrices also may be used to obtain a useful form for certain matrices, known as *LU-factorization*. In LU-factorization, a square matrix is written as the product of a lower triangular matrix and an upper triangular matrix. In the 3×3 case this looks like

$$A = \begin{bmatrix} l_{11} & 0 & 0 \\ l_{21} & l_{22} & 0 \\ l_{31} & l_{32} & l_{33} \end{bmatrix} \begin{bmatrix} u_{11} & u_{12} & u_{13} \\ 0 & u_{22} & u_{23} \\ 0 & 0 & u_{33} \end{bmatrix}$$

Upper and lower triangular matrices are particularly simple to work with for many reasons that will become clear later. In order to find an LU-factorization of a matrix $\mathbf{A}$, when it exists, one starts with the augmented matrix $\mathbf{I} \mid \mathbf{A}$ and uses elementary row operations to reduce the right side to upper triangular form (rather than going all the way to reduced echelon form). But here is the catch. *One may only use elementary row operations that involve adding a scalar multiple of one row to a row that is below it;* according to Exercise 2.13, such row operations correspond to *lower triangular* elementary matrices. When you are finished, the augmented matrix is

$$\mathbf{E}_n \cdots \mathbf{E}_1 \mid \mathbf{E}_n \cdots \mathbf{E}_1 \mathbf{A}$$

According to Exercises 2.12 and 2.13, the left matrix is a product of lower triangular matrices, so it, and its inverse $\mathbf{L} = \mathbf{E}_1^{-1} \cdots \mathbf{E}_n^{-1}$ are lower triangular. Letting $\mathbf{U} := \mathbf{E}_n \cdots \mathbf{E}_1 \mathbf{A}$, we see that $\mathbf{LU} = \mathbf{A}$.

In the next MP, you will only be able to perform the row operation just mentioned.

MP 32 *LU-FACTORIZATION*

Exercise 2.18 *Explain why the following matrix cannot be put into upper triangular form via only row operations that add a scalar multiple of one row to a row below it.*

$$\begin{bmatrix} 0 & 2 & 1 \\ 1 & 0 & 0 \\ 0 & 1 & 1 \end{bmatrix}$$

While not every square matrix has an LU-factorization, there are variations in this form that involve row swaps, which result in "permuted" LU-factorizations. For example, if the top two rows in the above matrix are swapped, then the resulting matrix has an LU-factorization, which requires only a single row operation to obtain.

3

The Determinant

The determinant function is one of the most remarkable functions in mathematics. It assigns to any square matrix $\mathbf{M}$ a real number $\det \mathbf{M}$ with many useful properties. Unlike most functions you have encountered, you will not first see the determinant as a formula, but rather as defined according to its properties. In fact, long before you see anything like a "formula" for the determinant (only for very small matrices), you will already have a method to compute it and understand its most important properties.

3.1 Basic Properties and Elementary Matrices

The idea that students sometimes have–that functions are always just formulas–is an illusion. Let's consider one simple function: $f(\theta) = \sin \theta$. There is a $\sin \theta$ button on your calculator, and $\sin \theta$ can be computed with a computer, but is it really a "formula"? Think again about how it is defined: On the unit circle, starting at the point $(1, 0)$, go counterclockwise until you have subtended an arc of length θ. Then $\sin \theta$ is the signed height above the x-axis. That is, $\sin \theta$ is defined by a geometric process, not a formula. (Yes, $\sin \theta$ has a Taylor Expansion, which one might consider as a "formula", but that's a story for another course.) Many important functions in physics are defined as solutions to differential equations and have no known "formulas"; they must be understood either by their properties or numerical approximations.

Here is an outline of how we will understand the determinant.

1. State the properties of the determinant.

2. Use those properties to develop a method to compute it (using Gauss-Jordan elimination).

3. Show that the determinant is the unique function with those properties.

4. Introduce *cofactor expansion*.

DOI: 10.1201/9781003674368-3

In a later chapter we will:

5. Show that the determinant has an important property called *multilinearity*.

6. Use multilinearity and uniqueness to prove that the determinant can be computed via cofactor expansion.

Cofactor expansion is the traditional way to introduce the determinant to undergraduates. It is a somewhat complex iterative process that produces simple "formulas" for the determinant of very small matrices. However, as we will see later, cofactor expansion is essentially useless to actually compute (even with the fastest computers) determinants of matrices that are even modestly large. For those matrices, good old Gauss-Jordan elimination, which you have already learned, is vastly more efficient.

Here are the defining properties of the determinant:

D1 $\det \mathbf{I} = 1$, where $\mathbf{I}$ is the identity matrix of any size.

D2 If any row of $\mathbf{A}$ is multiplied by a scalar k, the determinant of the resulting matrix is $k \det \mathbf{A}$.

D3 If a scalar multiple of one row of $\mathbf{A}$ is added to another, then the determinant of the matrix is unchanged.

You cannot have helped noticing that the last two conditions correspond to two of the elementary row operations, R1 and R2. This will ultimately tell us how to compute it. One of the most amazing facts about the determinant is that it is uniquely characterized by these three properties. This means that the determinant is the *only* function defined on square matrices that satisfies D1-D3. In first-year calculus you have already seen some functions that are uniquely characterized by their properties. For example, you should have learned in calculus that the exponential function $f(x) = e^x$ is the *one and only real differentiable function* such that $f'(x) = f(x)$ and $f(0) = 1$. The function $f(x) = 0$ also satisfies the first property, so the condition $f(0) = 1$ is necessary for e^x to be uniquely characterized. This condition is a "normalization" condition, as is D1. More generally, the functions $f(x) = e^{kx}$, where k is a non-zero constant, are the only differentiable real functions such that $f'(x) = kf(x)$ and $f(0) = 1$. Put another way, exponential functions are the only non-zero functions whose rate of change is proportional to their value. This property is very common in nature–from rates of growth of bacteria (under ideal circumstances) to exponential decay of radioactive materials. Since the exponential functions are the unique functions with this property, any physical phenomenon having a rate of change proportional to its value *must* be modeled by an exponential function.

Exercise 3.1 *Prove that real linear functions are uniquely characterized by their linearity. That is, if $f : \mathbb{R} \to \mathbb{R}$ is a real function such that for all real numbers k, a, b, $f(ka + b) = kf(a) + f(b)$, then f must be a linear function $f(x) = mx$. Use the following steps:*

    1. Define $m := f(1)$.

    2. Show that $f(0) = 0$ using $k = 0$ in the linearity condition. Hint: write $0 = k \cdot 0$.

    3. For any x, write $x = 1 \cdot x + 0$, and use linearity to conclude that $f(x) = mx$.

Of course there is one more row operation that hasn't been mentioned yet: swapping rows. But you already proved in Exercise 2.15 that swapping rows can always be carried out by five steps using the other two operations.

Exercise 3.2 *Revisit your proof in Exercise 2.15 and use properties D1-D3 to see what happens to the determinant with each row operation. For example, the first row operation is to add one row to another–which doesn't change the determinant. In fact, the only time the determinant changes is when you multiply a row by a scalar, and that happens once.... This proves the following property.*

    D4 If two rows of a matrix $\mathbf{A}$ are swapped, then the determinant of the resulting matrix is $- \det \mathbf{A}$.

Our first step to understand the determinant function is to find determinants of elementary matrices. Importantly, we will do so only using the properties D1-D4, and without any general method (yet) to compute determinants. We will then use determinants of elementary matrices to devise our first method to compute determinants for general square matrices.

    Properties D2-D4 describe the three row operations in Gauss-Jordan elimination. Since elementary matrices, by definition, are obtained from the identity matrix by applying one of the three row operations, finding determinants of elementary matrices is particularly simple. First, by D1, the determinant of the identity matrix $\mathbf{I}$ *must be* 1. If $\mathbf{E}$ is an elementary matrix obtained by multiplying one row of $\mathbf{I}$ by a non-zero scalar k, then D2 tells us that $\det(\mathbf{E})$ *must be* $k \det(\mathbf{I}) = k$. If $\mathbf{E}$ is an elementary matrix that adds adds a scalar multiple of one row of $\mathbf{I}$ to another, then by D3, it *must be* true that $\det \mathbf{E} = 1$. Finally, if an elementary matrix $\mathbf{E}$ is obtained from $\mathbf{I}$ by swapping two rows. Then D4 tells us that $\det(\mathbf{E})$ *must be* $- \det \mathbf{I} = -1$.

    I have emphasized the words *"must be"* four times as a reminder that conditions D1-D4 give us *no choice* for what the determinants of elementary matrices must be. That is, the determinant of any elementary matrix is *uniquely determined* by the properties D1-D4. We will eventually show that this is true for the determinant of any square matrix.

    Note also that in each case we saw that

$$\det \mathbf{E} \neq 0 \text{ when } \mathbf{E} \text{ is elementary}$$

One of the most important properties of the determinant is that for square matrices $\mathbf{M}$ and $\mathbf{N}$ of the same size, $\det(\mathbf{MN}) = \det \mathbf{M} \det \mathbf{N}$. We will prove this as a general theorem below, but for now we can check this for products $\mathbf{EN}$, where $\mathbf{E}$ is an elementary matrix and $\mathbf{N}$ is a square matrix. First of all, remember that if $\mathbf{E}$ is an elementary matrix that is obtained from $\mathbf{I}$ by a particular row operation R, then $\mathbf{EN}$ is obtained from $\mathbf{N}$ by applying the row operation R to $\mathbf{N}$.

Suppose that $\mathbf{E}$ swaps two rows when you multiply by it on the left. From the above discussion, we already know that $\det \mathbf{E} = -1$, and by D4, swapping rows negates the determinant of $\mathbf{N}$. So we can directly calculate that

$$\det(\mathbf{EN}) = -\det \mathbf{N} = (-1)\det \mathbf{N} = \det \mathbf{E} \det \mathbf{N} \tag{3.1}$$

The next exercise asks you to check the remaining two types of elementary matrices.

Exercise 3.3 *Finish the proof that if $\mathbf{E}$ is an elementary matrix and $\mathbf{N}$ is a matrix of the same size, then*

$$\det(\mathbf{EN}) = \det \mathbf{E} \det \mathbf{N}$$

by checking the two remaining row operations: multiplying a row by a scalar and adding a scalar multiple of one row to another.

Now suppose that $\mathbf{E}$ is any elementary matrix, which we already know must be invertible. In fact, $\mathbf{E}^{-1}$ is also an elementary matrix—it corresponds to the row operation that "undoes" the row operation of $\mathbf{E}$. See Theorem 2.11 and Exercise 2.12. We can now can apply Formula (3.1) to the product $\mathbf{E}\left(\mathbf{E}^{-1}\right) = \mathbf{I}$ to obtain $\det \mathbf{E} \det(\mathbf{E}^{-1}) = \det \mathbf{I} = 1$. As we observed above, $\det \mathbf{E} \neq 0$, and therefore

$$\det(\mathbf{E}^{-1}) = \frac{1}{\det \mathbf{E}} \tag{3.2}$$

We now have a way to compute the determinant of *any* non-singular matrix. According the Theorem 2.13, any non-singular matrix $\mathbf{A}$ may be written as a product of elementary matrices: $\mathbf{A} = \mathbf{E}_1 \cdots \mathbf{E}_n$. Using what you proved in Exercise 3.3 repeatedly, we can compute as follows:

$$\det \mathbf{A} = \det(\mathbf{E}_1 \cdots \mathbf{E}_n) = \det \mathbf{E}_1 \det(\mathbf{E}_2 \cdots \mathbf{E}_n)$$

$$= \det \mathbf{E}_1 \det \mathbf{E}_2 \det(\mathbf{E}_3 \cdots \mathbf{E}_n) = \cdots = \det \mathbf{E}_1 \cdots \det \mathbf{E}_n$$

But there is a subtle point here. There are infinitely many possible ways to write a non-singular matrix as a product of elementary matrices. If I write $\mathbf{A} = \mathbf{F}_1 \cdots \mathbf{F}_k$, where the $\mathbf{F}_i$'s are different elementary matrices, do I still get the same answer for the product of their determinants? This type of question is fundamental in mathematics and science: If I define a function in a way that involves making some *choices*, how do I know that my function doesn't depend on those choices? Formally speaking, this question is referred to as whether or not the function is *well-defined*.

In this case we can clearly state the question. Suppose we have written a non-singular square matrix $\mathbf{A}$ as a product of elementary matrices in two ways: $\mathbf{A} = \mathbf{E}_1 \cdots \mathbf{E}_n$ and $\mathbf{A} = \mathbf{F}_1 \cdots \mathbf{F}_k$. Is it true that $\det \mathbf{E}_1 \cdots \det \mathbf{E}_n = \det \mathbf{F}_1 \cdots \det \mathbf{F}_k$? To prove this, note that by Theorem 1.3, $\mathbf{A}^{-1} = \mathbf{F}_k^{-1} \cdots \mathbf{F}_1^{-1}$ and therefore

$$(\mathbf{E}_1 \cdots \mathbf{E}_n)\left(\mathbf{F}_k^{-1} \cdots \mathbf{F}_1^{-1}\right) = \mathbf{A}\mathbf{A}^{-1} = \mathbf{I}$$

Since all of the matrices involved are now elementary, we can use Equation (3.1) repeatedly to obtain

$$(\det \mathbf{E}_1 \cdots \det \mathbf{E}_n)\left(\det \mathbf{F}_1^{-1} \cdots \det \mathbf{F}_k^{-1}\right)$$

$$= \det(\mathbf{E}_1 \cdots \mathbf{E}_n \mathbf{F}_k^{-1} \cdots \mathbf{F}_1^{-1}) = \det \mathbf{I} = 1$$

By Equation (3.2) we have

$$(\det \mathbf{E}_1 \cdots \det \mathbf{E}_n)\left(\frac{1}{\det \mathbf{F}_1} \cdots \frac{1}{\det \mathbf{F}_k}\right) = 1$$

and therefore

$$\det \mathbf{E}_1 \cdots \det \mathbf{E}_n = \det \mathbf{F}_1 \cdots \det \mathbf{F}_k$$

The next theorem summarizes these calculations.

Theorem 3.1 *If $\mathbf{A}$ is a non-singular matrix then no matter how we write $\mathbf{A} = \mathbf{E}_1 \cdots \mathbf{E}_n$, as a product of elementary matrices,*

$$\det \mathbf{A} = \det \mathbf{E}_1 \cdots \det \mathbf{E}_n$$

and in particular $\det \mathbf{A} \neq 0$.

Now, as a practical matter, how do we compute the determinant of a non-singular matrix? As you remember, multiplication by an elementary matrix (on the left) corresponds to carrying out one of the row operations in Gauss-Jordan elimination. This means that we may compute the determinant of a *non-singular* matrix in the following way. Reduce the matrix to reduced echelon form via row operations, and keep track of what happens to the determinant with each row operation. For example, if our first row operation on a matrix $\mathbf{A}$ is multiplying some row by 3, then by D2, we know that the matrix $\mathbf{B}$ that we obtain after this operation satisfies $\det \mathbf{B} = 3 \det \mathbf{A}$. If our next row operation is to swap two rows, then the matrix $\mathbf{C}$ that we obtain satisfies

$$\det \mathbf{C} = (-1)\det \mathbf{B} = (-1)(3)\det \mathbf{A}$$

If our next step is to add a scalar multiple of one row to another to obtain a new matrix $\mathbf{D}$, then by D3, the determinant doesn't change. This corresponds to multiplying by 1 and we have the new equation:

$$\det \mathbf{D} = (1)\det \mathbf{C} = (1)(-1)(3)\det \mathbf{A}$$

Suppose that $\mathbf{D}$ is the reduced echelon form of $\mathbf{A}$. Since $\mathbf{A}$ is non-singular, this means that $\mathbf{D} = \mathbf{I}$ by Theorem 2.13. Since $\det \mathbf{I} = 1$, we get

$$1 = (-3) \det \mathbf{A} \text{ or } \det \mathbf{A} = \frac{-1}{3}.$$

To summarize: the general process to find the determinant of a non-singular matrix $\mathbf{A}$ is to use row operations to obtain the reduced echelon form, and keep track of the (non-zero) constants c_i that result from each row operation. Once the matrix is reduced to the identity, multiply all those constants together and invert the product to get the determinant. That is $\det \mathbf{A} = \frac{1}{c_1 c_2 \cdots c_m}$, where m is the number of row operations used to reduce $\mathbf{A}$ to $\mathbf{I}$.

MP 33 *DETERMINANT NON-SINGULAR*

In the previous MP, the matrix always reduced to the identity matrix because the original matrix was non-singular–and the resulting determinant was non-zero because all of the scalars were non-zero. But what if we want to compute the determinant of a singular matrix? According to Theorem 2.13, the reduced echelon form of a singular matrix $\mathbf{A}$ has at least one row of 0s. This fact now tells us what $\det \mathbf{A}$ *must be* 0, according to the following property that is a consequence of D1-D4.

Theorem 3.2 *If a matrix $\mathbf{A}$ has a row of 0s then $\det \mathbf{A} = 0$.*

We will check why this theorem is true for a 3×3 matrix $\mathbf{A} = \begin{bmatrix} a & b & c \\ d & e & f \\ 0 & 0 & 0 \end{bmatrix}$ by multiplying the bottom row by 2 (any scalar except 1 and 0 will work). For the third "=" below we use D2:

$$\det \mathbf{A} = \det \begin{bmatrix} a & b & c \\ d & e & f \\ 0 & 0 & 0 \end{bmatrix} = \det \begin{bmatrix} a & b & c \\ d & e & f \\ 2 \cdot 0 & 2 \cdot 0 & 2 \cdot 0 \end{bmatrix}$$

$$= 2 \det \begin{bmatrix} a & b & c \\ d & e & f \\ 0 & 0 & 0 \end{bmatrix} = 2 \det \mathbf{A}$$

That is, $\det \mathbf{A} = 2 \det \mathbf{A}$. But $\det \mathbf{A}$ is a real number, and the only real number that is equal to twice itself is 0.

Exercise 3.4 *Write down the proof of Theorem 3.2 for a square matrix of any size that has a row of 0s. You may simply mimic the above proof, but refer* *to the ith row in the following matrix:*

$$\begin{bmatrix} a_{11} & \cdots & a_{1j} & \cdots & a_{1n} \\ \vdots & \vdots & \vdots & \vdots & \vdots \\ 0 & \cdots & 0 & \cdots & 0 \\ \vdots & \vdots & \vdots & \vdots & \vdots \\ a_{n1} & \cdots & a_{nj} & \cdots & a_{nn} \end{bmatrix}$$

where the marked row is labelled i.

From this discussion we obtain yet another equivalent condition for non-singular matrices:

Theorem 3.3 *A square matrix* $\mathbf{A}$ *is non-singular if and only if* $\det \mathbf{A} \neq 0$.

Proof. If $\mathbf{A}$ is non-singular then by Theorem 3.1 we know that $\det \mathbf{A} \neq 0$. If $\mathbf{A}$ is singular, then $\mathbf{A} = \mathbf{E}_1 \cdots \mathbf{E}_n \mathbf{S}$, where $\mathbf{S}$ is the reduced echelon form of $\mathbf{A}$, and must have at least one row of 0s by Theorem 2.13. From Theorem 3.2 we know this means $\det \mathbf{A} = 0$. ∎

Remark 3.1 *As a reminder, the next theorem is extremely important (and it was, in the end, a lot work to get here!). One practical reason for its importance is that you might have designed an algorithm to compute determinants that you believe can be coded into very efficient software. But how do you know for sure that it really computes the determinant? One way to check this is simply to verify that the number produced by your algorithm satisfies D1-D3. Uniqueness is a concept that is deeply embedded in all forms of science, from classification of plants and animals to solutions of differential equations, and you should always remember and appreciate uniqueness theorems.*

Theorem 3.4 *The determinant is the unique function defined on square matrices that satisfies conditions D1-D3.*

Proof. Let D be an function satisfying those three properties. We used only properties D1-D3 to prove D4, so D4 must also be satisfied by D. We only used D1-D4 to show Theorem 3.1, which means that if $\mathbf{A}$ is non-singular, $D(\mathbf{A})$ *must be* $\det \mathbf{E}_1 \cdots \det \mathbf{E}_n = \det \mathbf{A}$, when we write $\mathbf{A} = \mathbf{E}_1 \cdots \mathbf{E}_n$ as a product of elementary matrices. That is, if $\mathbf{A}$ is non-singular, $D(\mathbf{A}) = \det \mathbf{A}$. On the other hand, we only used properties D1-D4 to prove Theorem 3.3, which means that when $\mathbf{A}$ is singular, $D(\mathbf{A}) = 0 = \det \mathbf{A}$. ∎

In the next MP, you will no longer have to enter the constants corresponding to each row operation–the MP will do that for you. And in each case the matrix is singular, so the goal is simply to obtain a row of 0s, at which point you are finished, and will enter 0 for the determinant.

MP 34 *DETERMINANT SINGULAR*

As promised earlier, we can now prove the following important theorem:

Theorem 3.5 *If* **A** *and* **B** *are square matrices of the same size, then* $\det(\mathbf{AB}) = \det\mathbf{A}\det\mathbf{B}$.

Proof. We may write $\mathbf{R} = \mathbf{E}_1\cdots\mathbf{E}_n\mathbf{A}$, where $\mathbf{R}$ is the reduced echelon form of $\mathbf{A}$ and the matrices $\mathbf{E}_i$ are elementary. Then

$$\mathbf{AB} = \mathbf{E}_n^{-1}\cdots\mathbf{E}_1^{-1}\mathbf{RB}$$

By Exercise 3.1,

$$\det(\mathbf{AB}) = \det\mathbf{E}_n^{-1}\cdots\det\mathbf{E}_1^{-1}\det(\mathbf{RB})$$

If $\mathbf{A}$ is non-singular then $\mathbf{R} = \mathbf{I}$ and

$$\det(\mathbf{AB}) = \det\mathbf{E}_n^{-1}\cdots\det\mathbf{E}_1^{-1}\det(\mathbf{B}) = \det(\mathbf{A})\det(\mathbf{B})$$

If $\mathbf{A}$ is singular, then $\det\mathbf{A} = 0$ and $\mathbf{R}$ has a row of 0s, say $\mathbf{R}_i = (0,0,\ldots,0)$. The entries in the ith row of $\mathbf{RB}$ are $(\mathbf{RB})_{ij} = \mathbf{R}_i \cdot \mathbf{B}^j = \mathbf{0} \cdot \mathbf{B}^j = 0$. In other words, the ith row of $\mathbf{RB}$ is also $\mathbf{0}$. By Theorem 3.2, $\det(\mathbf{RB}) = 0$, and putting this all together we see that

$$\det(\mathbf{AB}) = \det\mathbf{E}_1^{-1}\cdots\det\mathbf{E}_n^{-1}\det(\mathbf{RB}) = 0 = \det\mathbf{A}\det\mathbf{B}$$

∎

Exercise 3.5 *Use uniqueness to prove that for a 1×1 matrix $\mathbf{M} = [a]$, $\det(\mathbf{M}) = a$. To do so, define $D([a]) = a$. Check that the function D satisfies D1, D2, and D3. Then by uniqueness, it must be true that $D([a]) = a$.*

Example 3.1 *For a 2×2 matrix $\mathbf{M} = \begin{bmatrix} a & b \\ c & d \end{bmatrix}$, we define $D(\mathbf{M}) = ad - bc$.*

Then $D\left(\begin{bmatrix} 1 & 0 \\ 0 & 1 \end{bmatrix}\right) = 1 - 0 = 1$, *so D1 is satisfied. As for D2, the proof is basically the same for either row, so we check only the multiplication of the top row by a scalar:*

$$D\left(\begin{bmatrix} ka & kb \\ c & d \end{bmatrix}\right) = kad - kbc = k(ad - bc) = kD\left(\begin{bmatrix} ka & kb \\ c & d \end{bmatrix}\right)$$

Together with the next exercise, we have verified that this function D satisfies D1-D3. By uniqueness, this formula must give us the determinant of any 2×2 matrix.

Exercise 3.6 *By direct calculation, show that if $\mathbf{M} = \begin{bmatrix} a & b \\ c & d \end{bmatrix}$ and $\mathbf{N}$ is obtained from $\mathbf{M}$ by adding k times the top row of $\mathbf{M}$ to the bottom row of $\mathbf{M}$, then $D(\mathbf{N}) = D(\mathbf{N})\mathbf{M}$, where D is defined in Example 3.1.*

Exercise 3.7 *Show that* $\det(k\mathbf{M}) = k^2 \det \mathbf{M}$, *when* k *is a scalar and* $\mathbf{M}$ *is a* 2×2 *matrix. Use D2 to explain why, if* $\mathbf{M}$ *is an* $n \times n$ *matrix,* $\det(k\mathbf{M}) = k^n \det \mathbf{M}$.

Theorem 3.3 allows you to check whether or not a square matrix has an inverse by only calculating the determinant, which can be easier than trying (and possibly failing!) to compute its inverse. This is particularly simple for a 2×2 matrix $\mathbf{M} = \begin{bmatrix} a & b \\ c & d \end{bmatrix}$. As can be checked by direct calculation, if $\mathbf{M}$ has an inverse, meaning $\det \mathbf{M} \neq 0$, then $\mathbf{M}^{-1} = \frac{1}{\det \mathbf{M}} \begin{bmatrix} d & -b \\ -c & a \end{bmatrix}$. Since it is useful for pedagogical purposes to be able to invert 2×2 matrices quickly using this formula, in the next MP you will practice a little.

MP 35 *2X2 INVERSE*

We conclude this section with some additional basic theorems about determinants–and there will be more later! The next theorem is important for computational and other purposes. Among other things, it says that in order to compute the determinant of a matrix, you need only reduce it to upper triangular form–not all the way to reduced echelon form.

Theorem 3.6 *If* $\mathbf{M}$ *is upper triangular, then* $\det \mathbf{M}$ *is the product of its diagonal elements.*

The proof of this theorem will emerge from the next two MPs. In the next one, you will be restricted in what row operations you use.

MP 36 *DET UPPER TRIA NONSING*

In the prior MP, the upper triangular matrix was always non-singular. You first multiplied each row by the reciprocal of the leading entry, and the product of the resulting constants was just the product of the diagonal elements. After that, you could always finish the job using only the function to add a scalar multiple of one row to another, which does not change the determinant. In the next MP you will see what happens when there is a 0 on the diagonal of an upper triangular matrix.

MP 37 *DET UPPER TRIA SING*

Summarizing what you just verified: if there is a 0 on the diagonal of an upper triangular matrix, then the matrix can be reduced to a matrix with a row of 0s and therefore the determinant is 0. There–you have just done your first computer assisted proof! Congratulations!

You have also verified that it isn't necessary to reduce a matrix to reduced echelon form in order to compute the determinant–you can just stop when you reach upper triangular form, then multiply the diagonal elements and the inverse of the product of the scalars that you have accumulated. This, in turn, can be accomplished using only two operations: swapping rows (which negates the determinant) and adding a scalar multiple of one row to another, which doesn't impact the determinant. In the next MP you will practice this much faster method to find determinants. The only scalars to keep track of are the -1s that occur when you swap rows, which swaps -1 to 1, and vice verse. The MP will keep track of those for you.

MP 38 *FASTER DETERMINANT*

We already have one way to calculate the determinant (Gauss-Jordan elimination). If someone proposes a new way to calculate the determinant, they will have to prove that for every square matrix, of any size, their method gives the same number as the determinant. Doing so directly could be quite complicated. That's where uniqueness comes in handy. You have probably heard the saying: "If it walks like a duck and quacks like a duck, it is a duck." Like most such sayings, the statement isn't scientifically precise, but let's suppose for a moment that it is true. So you have and animal and you are wondering, is it a duck? Well, you don't have to do an expensive DNA test. You just have to answer two questions: Does it walk like a duck? Check. Does it quack like a duck? Check. It's a duck! In our case, to prove that a function is the determinant, we just need to show that *acts like the determinant* in three ways: D1-D3. Our final theorem of this section translates conditions D1-D3 into the language of elementary matrices. Since it is easier to code multiplication by elementary matrices than row operations, this theorem provides a useful way to understand properties D1-D3.

Theorem 3.7 *Suppose a function D assigns a real number $D(\mathbf{A})$ to any square matrix $\mathbf{A}$. If D satisfies the following two conditions, then $D(\mathbf{A}) = \det \mathbf{A}$ for every square matrix $\mathbf{A}$.*

> *E1* $D(\mathbf{I}) = 1$
>
> *E2 For every elementary matrix $\mathbf{E}$ corresponding to row operations D2 and D3, and square matrix of the same size, $D(\mathbf{EA}) = \det \mathbf{E} D(\mathbf{A})$.*

Proof. E1 is of course exactly the same as D1. Now if $\mathbf{E}$ is elementary of either of these two types, by properties E1 and E2,

$$D(\mathbf{E}) = D(\mathbf{EI}) = \det \mathbf{E} D(\mathbf{I}) = \det \mathbf{E}.$$

It remains to check D2-D3 for any square matrix $\mathbf{A}$. Suppose $\mathbf{B}$ is the matrix obtained from $\mathbf{A}$ by multiplying by $\mathbf{E}$. Then the property E2 shows that

$$D(\mathbf{B}) = D(\mathbf{EA}) = \det \mathbf{E} D(\mathbf{A})$$

If $\mathbf{E}$ multiplies a row by a non-zero scalar k, then we earlier saw that det $\mathbf{E} = k$, and the above equation becomes $D(\mathbf{B}) = kD(\mathbf{A})$, verifying D2. If $\mathbf{E}$ adds a scalar multiple of one row to another then det $\mathbf{E} = 1$ and likewise we have $D(\mathbf{B}) = D(\mathbf{A})$, verifying D3. ∎

Our first use of Theorem 3.7 looks innocuous enough because the transpose is one of the simplest matrix operations, so how hard can the proof be? But verifying this using uniqueness via D1-D3 is not going to be easy because conditions D2-D3 are about *row operations*, and when we take the transpose of a matrix, those row operations become *column operations*. In fact, one consequence of the next theorem is precisely that we may compute the determinant using column operations rather than row operations.

Theorem 3.8 *If $\mathbf{A}$ is a square matrix, then* $\det(\mathbf{A}^T) = \det(\mathbf{A})$.

Proof. We will define a new function: $D(\mathbf{A}) = \det(\mathbf{A}^T)$, and check the two conditions of Theorem 3.7. First,

$$D(\mathbf{I}) = \det(\mathbf{I}^T) = \det \mathbf{I} = 1$$

Now note that if $\mathbf{E}$ is any elementary matrix, then by Theorem 2.12, $\det(\mathbf{E}^T) = \det \mathbf{E}$. In fact, for the first and third types, $\mathbf{E} = \mathbf{E}^T$. For the third, $\det \mathbf{E} = 1 = \det \mathbf{E}^T$ because both $\mathbf{E}$ and $\mathbf{E}^T$ correspond to adding a scalar multiple of one row to another, which does not change the determinant. To finish the proof, we calculate:

$$D(\mathbf{EA}) = \det((\mathbf{EA})^T) = \det(\mathbf{A}^T\mathbf{E}^T) = \det\left(\mathbf{A}^T\right)\det(\mathbf{E}^T)$$

$$= \det(\mathbf{E}^T)\det\left(\mathbf{A}^T\right) = \det \mathbf{E} D(\mathbf{A})$$

That is, the second condition in Theorem 3.7 is satisfied. This means that $\det \mathbf{A} = D(\mathbf{A}) = \det(\mathbf{A}^T)$. ∎

Exercise 3.8 *Write down an explanation for each "=" in the last calculation above.*

From Theorem 3.8 we may now immediately conclude the following theorem, which we will often use without reference. It says basically that the impact of *column operations* on the determinant is exactly the same as the impact of row operations.

Theorem 3.9 *Properties D2-D4 are all true if "row" is replaced by "column" in each statement.*

We next have an expanded list of conditions equivalent to being non-singular, which combines Theorem 2.13 and Theorem 3.3:

Theorem 3.10 *The following are equivalent for a square matrix* $\mathbf{A}$:

1. $\mathbf{A}$ *is invertible.*

2. *Any equation* $\mathbf{AX} = \mathbf{B}$ *has a unique solution.*

3. *The homogeneous equation* $\mathbf{AX} = \mathbf{0}$ *has a unique solution.*

4. *The reduced echelon form of* $\mathbf{A}$ *is* $\mathbf{I}$.

5. $\mathbf{A}$ *is a product of elementary matrices.*

6. $\det \mathbf{A} \neq 0$

3.2   Cofactor Expansion

Because students are so used to working with formulas, many linear algebra texts define determinants via a method called "cofactor expansion", and require students to do tedious calculations with this method. Cofactor expansion does allow you to write down "formulas" in each dimension, but for matrices larger than 3×3 they rapidly become so monstrous that there is no point in even writing them down. For a 6×6 matrix the "formula" involves a sum of 720 numbers, each of which is the product of 6 numbers. As you will check later, the number of individual arithmetic calculations grows *factorially* as a function of the size of the matrix—vastly faster than exponentially. Therefore, as a method of calculation, cofactor expansion is literally useless for the kinds of large matrices that occur in modern science, engineering, finance, and other areas. Gauss-Jordan elimination,which you have already learned to use, is generally vastly more efficient.

When a textbook defines determinants via cofactor expansion, this simply introduces the opposite problem: proving that you can compute determinants with Gauss-Jordan elimination. In reality there is no work saved, and the important fact of uniqueness (which was easy to prove using D1-D3) often gets lost in the shuffle.

On the other hand, if it is so computationally useless, why study cofactor expansion at all? First, cofactor expansion is a cautionary tale: while it can provide a formula for the determinant that involves only repeated multiplication and addition, which might make it *naively* appealing to implement, it is generally impossible to use for many modern problems. The moral of the story is that, when it comes to applying mathematics to modern problems, the simplest solutions to state, derive and implement may not be the best from a practical standpoint. The cutting edge of science involves not just implementing old mathematical methods on faster machines, but also finding new (and often surprising) mathematical methods that can analyze the vast amounts of data that modern humanity creates and uses. This is why jobs in the mathematical sciences, especially applied mathematics, are among the fasted growing and highly paid in all of STEM.

A more positive reason to study cofactor expansion is that there are basic concepts, notation, and methods that can be introduced that are used in other important contexts. Therefore, your focus in this section should be not on *calculation* of determinants, but on the concepts underlying cofactor expansion. One of the most important of these concepts is that cofactor expansion is defined *iteratively*. Iterative definitions are closely related to the powerful method of *mathematical induction* and are very common in mathematics, science and engineering—especially computer science.

Since we don't know *a priori* that our calculation actually gives us the determinant, we will consider the construction of a function $D(\mathbf{A})$ as we did in the proof of Theorem 3.8, and we will show later that this function that we have created is equal to the determinant. For a 1×1 matrix $\mathbf{A} = [a]$ define $D(\mathbf{A}) = a$, which we saw previously is equal to $\det[a]$. This $n = 1$ case is called the *base case*. Next we suppose that we have defined $D(\mathbf{A})$ iteratively for any $n \times n$ matrix with $n \geq 1$. We will then show how to define D for $(n+1) \times (n+1)$ matrices, using the definition for $n \times n$ matrices. This is called the *iterative step*.

The iterative step is a bit complicated, and we need some new notation. Suppose $\mathbf{A} = [a_{ij}]$ is a square $(n+1) \times (n+1)$ matrix. The (i,j) *minor matrix* of $\mathbf{A}$ is the $n \times n$ matrix $\mathbf{M}_{ij}$ obtained from $\mathbf{A}$ by removing the ith row and the jth column of $\mathbf{A}$. Before going further, I will warn that there is **no consistency** among matrix algebra texts and online materials concerning the word "minor" in this context, which variously refers to either the minor matrix $\mathbf{M}_{ij}$ as defined above, or the determinant of $\mathbf{M}_{ij}$. Even more confusingly, sometimes a capital (but not bold) "M_{ij}" is used to denote the latter "minor", which, as a determinant, is a number, not a matrix. In the this text I will *only* refer to the "minor *matrix*" $\mathbf{M}_{ij}$, and will never use the word "minor" as a noun to describe anything.

The best way to understand minor matrices is to actually see them, which you will do in the next MP. Remember: to find the minor matrix $\mathbf{M}_{ij}$, eliminate the ith row and j^{th} column, and $\mathbf{M}_{ij}$ is what is left.

MP 39 *MINOR MATRICES*

Generally speaking, an $(n+1) \times (n+1)$ matrix $\mathbf{A}$ looks like the following:

$$A = \begin{bmatrix} a_{11} & a_{12} & \cdots & a_{1,n} & a_{1,n+1} \\ a_{21} & a_{22} & \cdots & a_{2,n} & a_{2,n+1} \\ \vdots & \vdots & \vdots & \vdots & \vdots \\ a_{n1} & a_{n2} & \cdots & a_{n,n} & a_{n,n+1} \\ a_{n+1,1} & a_{n+1,2} & \cdots & a_{n+1,n} & a_{n+1,n+1} \end{bmatrix}$$

Since $\mathbf{M}_{ij}$ is the matrix obtained when removing the ith row and j^{th} column from $\mathbf{A}$, $\mathbf{M}_{ij}$ is an $n \times n$ matrix. Therefore, by our iterative assumption, we in principle know how to compute $D(\mathbf{M}_{ij})$. We will describe cofactor expansion "along the first row", and define

$$D(\mathbf{A}) := a_{11}D(\mathbf{M}_{11}) - a_{12}D(\mathbf{M}_{12}) + \cdots \pm a_{1,n+1}D(\mathbf{M}_{1,n+1}) \qquad (3.3)$$

The $\pm$ in the last term depends on whether n is odd or even, in a way that we will see exactly below. While this seems very abstract, you should take a moment to read this carefully because these kinds of expressions are very common in mathematics, science and engineering when matrix algebra is involved. Visually speaking, we are going across the first row, starting with a_{11}. We mentally "cross out" row 1 and column 1 to see $\mathbf{M}_{11}$. Since $\mathbf{M}_{11}$ is an $n \times n$ matrix, by our iterative assumption we know how to calculate $D(\mathbf{M}_{11})$. The first summand in $D(\mathbf{A})$ is $a_{11}D(\mathbf{M}_{11})$. This sum is an *alternating sum*, something that occurs widely in nature. This means that for every other term, we insert a minus sign. For the second summand, we mentally cross out the top row and the second column to see $\mathbf{M}_{12}$, calculate $D(\mathbf{M}_{12})$, and multiply by $-a_{12}$, and so on. As for the last term, of course its sign depends on whether $n+1$ is odd ($+$) or even ($-$). It is useful to compress this sum using summation notation. The main issue is that we don't want to have separate sums for odd or even n. The convenient way to do this is to use powers of -1, since $(-1)^n$ is equal to -1 when n is odd and equal to 1 when n is even. That is,

$$D(\mathbf{A}) = (-1)^2 a_{11}D(\mathbf{M}_{11}) + (-1)^3 a_{12}D(\mathbf{M}_{12}) + \cdots \qquad (3.4)$$

$$\cdots + (-1)^{n+2} a_{1,n+1}D(\mathbf{M}_{1,n+1})$$

Take some time to read the above expression carefully and be sure that you understand *every part of every term*. One of the most common reasons that students fail at mathematics is that they do not learn the language of mathematics well enough. As a good reality check, find another student who knows or is learning linear algebra (even an imaginary student!), and explain the above expression to them in complete detail. For example, what is the "3" doing there and why is the power in the last term $n + 2$ rather than $n + 1$? If you cannot confidently explain the answers to questions like that then you are not ready to succeed in the rest of this course. Ask your instructor about this if you are at all unsure after working through it carefully.

Let's express this sum with summation notation:

$$D(\mathbf{A}) = \sum_{k=1}^{n+1} (-1)^{k+1} a_{1k} D(\mathbf{M}_{1k})$$

Again, you should look very carefully at the summands to make sure everything "adds up". Does k start and end with the right numbers? Do the various terms match what you see in Formula (3.4)? For example, the powers of -1 should go from 2 to $n + 2$, and yes, $k + 1$ goes from 2 to $n + 2$. If this seems complicated–just remember that *this is the way determinants are typically introduced in a course at this level!* Again, the calculation itself is not the important part–focus on understanding the expressions.

It is possible to do cofactor expansion across the jth row (not just the top row), by tweaking the same formula to obtain

$$\sum_{k=1}^{n+1} (-1)^{k+j} a_{jk} D(\mathbf{M}_{jk})$$

Likewise, the sum via cofactor expansion in the jth column can be written as follows:

$$D(\mathbf{A}) = \sum_{i=1}^{n+1} (-1)^{k+j} a_{kj} D(\mathbf{M}_{kj})$$

Again, you should take some time to be sure you can properly read these expressions. One of the goals for this course (and every mathematics course!) is for you to become proficient with the notation so that you can quickly read similar expressions that you see in the future.

We already know that $D([a]) = a$, which is also $\det[a]$. With this "base case" we will iteratively define $D(\mathbf{A})$ for $n+1$ matrices using $D(\mathbf{A})$ for $n \times n$ matrices. Let's see how this works for small matrices. Consider the 2×2 case, with matrix

$$\begin{bmatrix} a & b \\ c & d \end{bmatrix}$$

$\mathbf{M}_{11}$ is obtained by removing the top row and first column, leaving the minor matrix $[d]$, so the first term in the cofactor expansion is $a(-1)^2 d = ad$. Likewise the second term is $b(-1)^3 c = -bc$. So $D(\mathbf{A}) = ad - bc$, which as you will remember is the formula that we already derived for $\det \mathbf{A}$. Now we can take the iterative step to compute the D for a 3×3 matrix:

$$A = \begin{bmatrix} a & b & c \\ d & e & f \\ g & h & k \end{bmatrix}$$

Then $\mathbf{M}_{11} = \begin{bmatrix} e & f \\ h & k \end{bmatrix}$, which has determinant $ek - hf$, so the first term in the cofactor expansion across the top row is $a(ek - hf)$. The second term is $-b \det \begin{bmatrix} d & f \\ g & k \end{bmatrix} = -b(dk - gf)$, and you may check the last term to see that

$$D(\mathbf{A}) = a(ek - hf) - b(dk - gf) + c(dh - ge) \tag{3.5}$$

Again, it is very important for you to take the time to work this very carefully yourself so that you understand the process! Now you should see where things are going. For a 4×4 matrix we have four terms with alternating signs, each of which is an entry from the top row times the determinant of the corresponding minor matrix, which is a 3×3 matrix–for which we have Formula (3.5). This is getting complicated fast! How bad does it get? A simple measure of complexity in this case is the number $S(n)$ of summands in the formula, and the number of numbers $M(n)$ multiplied for each summand, for an $n \times n$ matrix. From the computations above we know that $S(1) = M(1) = 1$, $S(2) = 2$, $M(2) = 2$, $S(3) = 6$, and $M(3) = 3$. So far this doesn't look to bad, but the next exercise shows how bad $S(n)$ gets.

Exercise 3.9 *Explain why $S(4) = 24$ and $M(4) = 4$. Now explain why $S(n) = n!$ and $M(n) = n$.*

The problem with the factorial function is that it grows extremely fast. For example, according to some estimates of world computing power, if every computer on the planet were harnessed to compute the determinant of a relatively small 35×35 matrix using cofactor expansion, it would take thousands of years! So yes, strictly speaking there is a cofactor expansion "formula" for the determinant of any $n \times n$ matrix, but actually computing the determinant of even a 3×3 matrix by hand is immensely tedious, with the potential to make lots of small 5th-grade errors. The only ones that you should be reasonably be asked to do by hand have a lot of 0s in the row or column in which you do the cofactor expansion. And the only reason to ask you to do so is to give you an opportunity to interact with the summation notation. For example:

$$A = \begin{bmatrix} 1 & 1 & 3 & 2 \\ 2 & 0 & 0 & 1 \\ 0 & 0 & 0 & 4 \\ 0 & 1 & 0 & 1 \end{bmatrix}$$

There is a row with three 0s and a column with three 0s, and for either one, all but one term in the cofactor expansion vanish. Let's expand along the third column. The minor matrix $\mathbf{M}_{13}$ is $\begin{bmatrix} 2 & 0 & 1 \\ 0 & 0 & 4 \\ 0 & 1 & 1 \end{bmatrix}$. So the first and only summand in det $\mathbf{A}$ is $(3)(-1)^{1+3} = 3 \det \mathbf{M}_{13}$. We can find $\det \mathbf{M}_{13}$ by expanding across, say, its middle row. This row has two zeros, so the only non-zero summand in $\det \mathbf{M}_{13}$ is the last one. Therefore, $\det \mathbf{M}_{13} = 4(-1)^{2+3} \det \begin{bmatrix} 2 & 0 \\ 0 & 1 \end{bmatrix} = -8$. Therefore $\det \mathbf{A} = 3(-8) = -24$. Now take a moment to think about how rapidly you could use row operations in the Gauss-Jordan MP to reduce even this "easy" matrix to upper triangular form to compute its determinant.

Exercise 3.10 *Here are some 0-heavy matrices. Calculate the determinant values. The only point of this exercise is for you to think about a good strategy. Otherwise, I'm sorry to make you do it! Again, you could get the answer from a computer, but that's completely missing the point!*

1. $\det \begin{bmatrix} 0 & 1 & -1 & 2 \\ 0 & 2 & 0 & 0 \\ 1 & 0 & 1 & 0 \\ 1 & 1 & 0 & 3 \end{bmatrix} = 2$

2. $\det \begin{bmatrix} 0 & 1 & -1 & -2 \\ 0 & 2 & 0 & -1 \\ 1 & 0 & 1 & 0 \\ 0 & 0 & 0 & 3 \end{bmatrix} = 6$

3. $\det \begin{bmatrix} 0 & 1 & -1 & 2 \\ 0 & 2 & -2 & 0 \\ 0 & 2 & 0 & 0 \\ -1 & 1 & 0 & 3 \end{bmatrix} = 8$

For the next MP, the software will do most of the work. Your only job will be to enter the constant terms (with the proper signs!) and select the minor matrix, and the MP will compute the determinant of the minor matrix for you.

MP 40 *COFACTOR EXPANSION*

At this point we only know that the cofactor expansion $D(\mathbf{A})$ gives us $\det \mathbf{A}$ for 1×1 and 2×2 matrices. We will defer to the next chapter the proof that $D(\mathbf{A}) = \det \mathbf{A}$ for any square matrix, because the proof requires some additional techniques that you have yet to learn. You should take a moment to appreciate how different these two approaches are to compute the determinant, and how, at this point, there is no obvious reason to believe that they are the same, or an obvious way to prove it. This is a perfect example of how "knowledge is power". Someone who only knows how to compute determinants via cofactor expansion will literally have no way to compute the determinant for large matrices. Someone who knows that you can efficiently compute determinants using Gauss-Jordan elimination will have a huge advantage. Applied mathematicians work with scientists, engineers and finance experts to develop new (and often unexpected) ways to improve our understanding of the world and the ways in which to compute what is needed. Any scientist or engineer who intends to work at the cutting edge of their field needs to understand mathematics well enough to be able to collaborate with mathematicians who are discovering new mathematics that advances the underlying science.

I will conclude this section by barely mentioning a staple of sophomore-level linear algebra courses known as Cramer's Rule, which is a computationally absurd method involving determinants to find inverses for matrices (and hence for solving systems of linear equations). It is hard to imagine a modern purpose for this topic in a basic linear algebra course—its former purpose seems to have been honing students' skills in adding and multiplying numbers with pencil and paper. Therefore I will not expend any more effort on this topic in this text. Cramer's Rule does have some theoretical purposes well beyond the scope of this course; the interested student may wish to read about the subject in more advanced texts or online.

4

Euclidean Vector Spaces

In this chapter we introduce Euclidean vector spaces. You have already learned about the Euclidean plane back in high school, which we will denote by $\mathbb{R}^2$. As a reminder, $\mathbb{R}^2$ is the set of all ordered pairs (x, y) where x and y are real numbers. As by now you know, the physical world, for many purposes, can be modeled with Euclidean Space $\mathbb{R}^3$, which consists of ordered triples (x, y, z) of real numbers. There is no reason to stop with $n = 3$, however. Modern problems often involve many variables, and the higher-dimensional Euclidean spaces $\mathbb{R}^n$ are extraordinarily useful in modern science. We will say more about "dimension" later, but for the moment we will refer to $\mathbb{R}^n$ as "n-dimensional Euclidean space" and you can think of it as the realm of n independent variables. The points in $\mathbb{R}^n$ consist of ordered n-tuples $(x_1, \ldots, x_n)$ of real numbers. The names of the variables don't matter from a mathematical standpoint, but there are some conventions that humans use, especially when variables refer to standard quantities such as position (x, y, z) in space ($\mathbb{R}^3$), and time t.

4.1 Vectors in Dimension 2

As it turns out, from the standpoint of linear algebra, much of the geometric intuition that you need to understand vectors and their applications may be found in plane geometry. Therefore, we will carefully work out the relationship between linear algebra and geometry in the setting of $\mathbb{R}^2$ and then extend the ideas to higher dimensions in the next section.

Ordered pairs contain the exact same information as 1×2 *row vectors* $\begin{bmatrix} x & y \end{bmatrix}$ or 2×1 *column vectors* $\begin{bmatrix} x \\ y \end{bmatrix}$—that is, two real numbers in a specified order. An ordered pair of numbers can be used to represent many things, and these different notations are used in different contexts. In fact, plane points and plane vectors are often both denoted by an ordered pair (x, y). Vectors (but never points) are also represented as row or column vectors according to the context and purpose. There are some notational differences that will help you distinguish between points and vectors. In *Interactive Linear Algebra*, vectors

DOI: 10.1201/9781003674368-4

will always be denoted with bold small letters, e.g. $\mathbf{v} = (1, 2)$, while points will always be denoted by capital non-bold letters, like $P = (1, 2)$ or equivalently $P(1, 2)$. For the point $P(1, 2)$, the numbers 1 and 2 are called "coordinates" as you have always done, but for the vector $\mathbf{v} = (1, 2)$, the numbers are usually called "components"—although "coordinates" is also used.

Another aspect of vectors that may seem confusing at first, is that vectors may be considered as being "placed at a point", as we will now describe. At any point P in the plane, a vector "at P" may be geometrically described as an arrow with its "base" at P and its "tip" at some other point. If vector $\mathbf{v} = (v_1, v_2)$ is based at a point $P = (x, y)$ then $\mathbf{v}$ is represented as an arrow with base at P and tip at the point $P(x + v_1, y + v_2)$. Put another way, to geometrically represent $\mathbf{v}$ at P, place the base of $\mathbf{v}$ at P, then move in the x-direction an amount v_1, and in the y-direction an amount v_2. When we refer to "moving in the x-direction amount v_1", this means that we move to the right if v_1 is positive, to the left if v_2 is negative, and do not move at all of $v_1 = 0$. Similarly, we move up, down, or not at all, depending on the sign of v_2. You will practice plotting vectors in the next MP:

MP 41 *PLOTTING VECTORS*

We will formally define the "length" of a vector later, but you can see geometrically how to compute this for plane vectors using the Pythagorean Theorem. That is, when the vector (v_1, v_2) is plotted at a point P, then the distance from the base of the vector at $P = (x, y)$ and the tip of the vector $(x + v_1, y + v_2)$ is

$$\sqrt{((x + v_1) - x)^2 + ((y + v_2) - y)^2} = \sqrt{v_1^2 + v_2^2}$$

The concept of a vector at a point is best introduced by a concrete application, and we will use one that you may have seen in calculus, namely *velocity vectors*. Suppose an object is moving in the plane according to a differentiable *position function* $p(t) = (x(t), y(t))$. That is, at any time t, the object is located at the point with x-coordinate $x(t)$ and y-coordinate $y(t)$, and these two functions are differentiable real functions. For example, at $t = 2$, the object is located at the point $(x(2), y(2))$ in the plane; if $p(t) = (t, t^2)$ then $p(2) = (2, 4)$. The velocity vector represents the change in position at the instant t. It is computed using the derivative, as $\mathbf{v}(t) = p'(t) = (x'(t), y'(t))$. In the present example, $\mathbf{v}(t) = p'(t) = (1, 2t)$. (One may go further than this, which you may have seen in a calculus or physics class, defining the acceleration vector to be $\mathbf{a}(t) = p''(t) = (x''(t), y''(t))$, but we will not need to do so in this text.) Note that for any t, $p'(t)$ is an ordered pair of numbers, but we interpret it as a vector, not a point, and geometrically $\mathbf{v}(t)$ is plotted with its base at the point $p(t) = (x(t), y(t))$.

Imagine a car driving on a flat plane a night. The velocity vector of the car points in the direction of the headlights, which changes as the car navigates curves. The length of the velocity vector, which is a number, represents speed of the car. These two aspects of a vector—its direction and length, completely characterize the vector, as we will see more formally later. How are the "speed" and "direction" calculated from the ordered pair $\mathbf{v}(t) = (x'(t), y'(t))$? Let's consider a relatively basic example. Let $p(t) = (x(t), y(t)) = (\sin t, \cos t)$, which represents motion around the unit circle. The motion starts at the point $P(0, 1)$ for $t = 0$ and then travels clockwise around the circle as the time t increases, and counterclockwise as t decreases. You can see that this path lies on the unit circle, because

$$x(t)^2 + y(t)^2 = \sin^2 t + \cos^2 t = 1$$

which is the equation that all points (x, y) on the unit circle satisfy. The velocity vector at time t is

$$\mathbf{v}(t) = (\cos t, -\sin t)$$

How do we find the magnitude and direction from the components of the vector? Suppose we are at time $t = \frac{\pi}{4}$. Then we are at position

$$p\left(\frac{\pi}{4}\right) = \left(\sin\frac{\pi}{4}, \cos\frac{\pi}{4}\right) = \left(\frac{\sqrt{2}}{2}, \frac{\sqrt{2}}{2}\right)$$

on the unit circle. Since

$$\mathbf{v}\left(\frac{\pi}{2}\right) = p'\left(\frac{\pi}{4}\right) = \left(\cos\frac{\pi}{4}, -\sin\frac{\pi}{4}\right) = \left(\frac{\sqrt{2}}{2}, -\frac{\sqrt{2}}{2}\right)$$

the interpretation of the velocity vector is that at the instant $t = \frac{\pi}{4}$, if the motion were suddenly to become straight, then in one unit of time the object would move $\frac{\sqrt{2}}{2}$ in the x-direction and $-\frac{\sqrt{2}}{2}$ in the y-direction. This can be visualized as an arrow on the unit circle that is tangent to the circle. Again, imagine a car traveling around the unit circle. Its headlights would point in the direction of the tangent vector at that instant. The length of the arrow is the distance that would be covered by the car if you let go of the steering wheel at that instant and the car continued straight–and would also be the speed that you see on the speedometer (with appropriate units, of course!). The speed $s(t)$ can be calculated using the Pythagorean Theorem. For example, at $t = \frac{\pi}{4}$, the motion is $\frac{\sqrt{2}}{2}$ in the x-direction and $-\frac{\sqrt{2}}{2}$ in the y-direction, and therefore $s(\frac{\pi}{4}) = \sqrt{\left(\frac{\sqrt{2}}{2}\right)^2 + \left(\frac{\sqrt{2}}{2}\right)^2} = 1$. In fact, the motion described by this equation always has speed 1 because

$$s(t) = \sqrt{(\cos t)^2 + (-\sin t)^2} = \sqrt{\cos^2 t + \sin^2 t} = 1$$

But the *direction* of travel is *not* constant—it is a vector that changes continuously as the object travels around the circle, and always points in the tangent direction.

Formally speaking, a "direction" is a vector of length 1, also known as a *unit vector*. The idea here is that your "direction" should not depend on how fast you are going, so the easiest thing is to just take "direction" to have length 1. The exception is when your speed is 0, in which case, strictly speaking, you have no direction.

If we change the position function to $p(t) = (\sin(t^2), \cos(t^2))$ then the travel is still on the unit circle because $\sin^2(t^2) + \cos^2(t^2)$ is still 1, but the movement, and therefore the velocity, is more complicated. Using the chain rule in each component,

$$\mathbf{v}(t) = (2t \cos(t^2), -2t \sin(t^2))$$

Now the speed is

$$s(t) = \sqrt{4t^2 \sin^2(t^2) + 4t^2 \cos^2(t^2)} = 2\,|t|\,\sqrt{\sin^2(t^2) + \cos^2(t^2)} = 2\,|t|$$

Therefore the speed is not always 1 and is even 0 when $t = 0$. At the instant $t = 0$, the motion has "stopped" and has no direction.

To summarize, a plane vector $\mathbf{v} = (v_1, v_2)$ at some point $P(x, y)$ may be represented *algebraically*, as an ordered pair of numbers, and *geometrically* as an arrow based at the point P, which has a length or *magnitude* and a *direction*, as long as $\mathbf{v} \neq (0,0)$. When $\mathbf{v} = (0,0)$ we will also write $\mathbf{v} = \mathbf{0}$ (note the bold $\mathbf{0}$ to distinguish the vector from the scalar 0), in which case $\mathbf{v}$ *has no direction*. The geometric and algebraic interpretations are both useful, and we will discuss them in tandem. In fact, the general interplay between algebra and geometry is one of the deepest and most useful relationships in all of mathematics, with important applications in the sciences and engineering.

As mentioned earlier, vectors can expressed as row or column vectors, which are also matrices–and algebraically we may perform the same operations on them that we can with matrices. For example, if $\mathbf{v} = (v_1, v_2)$ and $\mathbf{w} = (w_1, w_2)$ then $\mathbf{v} + \mathbf{w} = (v_1 + w_1, v_2 + w_2)$. What is the *geometric* expression of vector addition in the plane? In MP 41, you saw that to plot a vector placed at a point P in the plane, you move in the x direction by the quantity v_1 in the first component, and in the y-direction by the quantity v_2 in the second component. Considering the sum $\mathbf{v} + \mathbf{w} = (v_1 + w_1, v_2 + w_2)$, you move in the x-direction by $v_1 + w_1$ and in the y-direction $v_2 + w_2$. You may do this in increments: first move in the x direction by v_1 and the y-direction by v_2. This places you at the tip of $\mathbf{v}$. From that point you move w_1 in the direction x and w_2 in direction y. But this is the same as plotting $\mathbf{w}$ based at the tip of $\mathbf{v}$. Therefore, geometrically speaking, you may plot $\mathbf{v} + \mathbf{w}$ at a point P by first plotting $\mathbf{v}$ at P and then plotting $\mathbf{w}$ at the tip of $\mathbf{v}$. You will practice this in the next MP.

MP 42 *ADDING VECTORS*

As you know, matrices have an operation consisting of multiplication by scalars. Considering vectors as matrices, we may use this operation on vectors as well. That is, if $\mathbf{v} = (v_1, v_2)$ then $k\mathbf{v} = (kv_1, kv_2)$. Let's now consider the geometric effect of scalar multiplication, say $2\mathbf{v}$. To plot the vector $2\mathbf{v}$ at P, you start at P and move $2v_1$ in the direction of x, and $2v_2$ in the direction of y. The vector obtained points in the same direction as $\mathbf{v}$, but with twice the length. It should be intuitively clear that in general, if $k > 0$, $k\mathbf{v}$ is the vector that points in the same direction as $\mathbf{v}$, but with length multiplied by k. If $k = 0$ then $k\mathbf{v} = (0,0) = \mathbf{0}$, which has no direction. If $k = -1$, then we are negating the components, and $-\mathbf{v}$ points in the opposite direction as $\mathbf{v}$ with the same length. More generally, if $k < 0$, $k\mathbf{v}$ is obtained by "flipping" $\mathbf{v}$ to the opposite direction $(-\mathbf{v})$ and multiplying its length by $|k|$.

MP 43 *SCALAR MULTIPLICATION OF VECTORS*

As with matrices and real numbers, subtraction of vectors $\mathbf{v} - \mathbf{w}$ is simply a familiar notation for $\mathbf{v} + (-\mathbf{w})$. That is, geometrically speaking, you flip $\mathbf{w}$ to $-\mathbf{w}$ and plot it at the tip of $\mathbf{v}$.

MP 44 *SUBTRACTION OF VECTORS*

As you see have seen, to find the sum $\mathbf{v} + \mathbf{w}$ of two vectors geometrically, translate $\mathbf{v}$ to the tip of $\mathbf{w}$: then $\mathbf{v} + \mathbf{w}$ is the vector from the base of $\mathbf{w}$ to the tip of the translated vector $\mathbf{v}$. We know that addition is commutative because matrix addition is commutative, i.e. $\mathbf{v} + \mathbf{w} = \mathbf{w} + \mathbf{v}$, but you can also see this geometrically because the individual vectors $\mathbf{v}$ and $\mathbf{w}$ and their translates form a parallelogram, with $\mathbf{v} + \mathbf{w} = \mathbf{w} + \mathbf{v}$ as the diagonal. We will refer to this parallelogram as the *parallelogram spanned by* $\mathbf{v}$ and $\mathbf{w}$. This parallelogram plays an important role in the relationship between linear algebra and geometry, and you will see it visually in the next MP.

MP 45 *COMMUTATIVE LAW OF VECTORS*

In the next MP you will visualize the associative law for plane vectors.

MP 46 *ASSOCIATIVE LAW OF VECTORS*

Since vectors may be considered as matrices, you are probably already wondering about how matrix multiplication works for vectors. You should recall that when we introduced the subject of matrix multiplication, we described the "dot product" of row i of matrix $\mathbf{M}$ with column j of matrix $\mathbf{N}$, which gave us the i, j entry of the product $(\mathbf{MN})_{ij}$. Put another way, to take the dot product $\mathbf{v} \cdot \mathbf{w}$ of vectors $\mathbf{v}$ and $\mathbf{w}$ we can treat $\mathbf{v}$ as a row vector $\begin{bmatrix} v_1 & v_2 \end{bmatrix}$ and $\mathbf{w}$ as a column vector $\begin{bmatrix} w_1 \\ w_2 \end{bmatrix}$ perform matrix multiplication

$\begin{bmatrix} v_1 & v_2 \end{bmatrix} \begin{bmatrix} w_1 \\ w_2 \end{bmatrix} = [v_1 w_1 + v_2 w_2]$. The single number in this 1×1 matrix is the *dot product* of the vectors, $\mathbf{v} \cdot \mathbf{w} = v_1 w_1 + v_2 w_2$. Why didn't I just write that formula out to begin with, as you may have seen in calculus, and skip the step of considering the vectors as matrices? The advantage of describing the dot product via matrix multiplication is that we can use everything that we proved about matrix multiplication without doing any more work. For example, from the distributive law for matrix multiplication and addition, we know that the distributive law holds for the dot product:

$$\mathbf{v} \cdot (\mathbf{u} + \mathbf{w}) = \mathbf{v} \cdot \mathbf{u} + \mathbf{v} \cdot \mathbf{w}$$

and scalar multiplication satisfies $k(\mathbf{u} \cdot \mathbf{v}) = (k\mathbf{u}) \cdot \mathbf{v} = \mathbf{u} \cdot (k\mathbf{v})$. If you need a refresher on dot products, you are encouraged to revisit MP 12.

There is an important difference between the dot product and more general matrix multiplication. As you recall, matrix multiplication is generally *not* commutative. That is, there are many (square) matrices $\mathbf{M}$ and $\mathbf{N}$ such that $\mathbf{MN} \neq \mathbf{NM}$. However, as can be seen by direct computation the dot product is always commutative:

$$\mathbf{v} \cdot \mathbf{w} = v_1 w_1 + v_2 w_2 = w_1 v_1 + w_2 v_2 = \mathbf{w} \cdot \mathbf{v}$$

Finally, there can be no notion of "inverses" of vectors, since they are not square. So for example, if $\mathbf{v} \cdot \mathbf{w} = \mathbf{u} \cdot \mathbf{w}$ then there is no "cancellation" of the $\mathbf{w}$ on each side of this equation and we cannot conclude that $\mathbf{u} = \mathbf{v}$. In fact, as you will see in the next exercise, that conclusion is false even when $\mathbf{w} \neq (0,0)$.

Exercise 4.1 *Find vectors* $\mathbf{u}, \mathbf{v}, \mathbf{w}$ *with* $\mathbf{w} \neq (0,0)$ *such that* $\mathbf{v} \cdot \mathbf{w} = \mathbf{u} \cdot \mathbf{w} = 0$ *but* $\mathbf{u} \neq \mathbf{v}$.

Although the next computation is simply a result of the distributive law, and it should look entirely familiar from algebra of the real numbers, it is worth writing it out to see the role of the commutative law for the dot product, and as a lesson in remembering what is a scalar and what is a vector. In fact, $\mathbf{u} \cdot \mathbf{v}$ is by definition a scalar—and for this reason it is often called the *scalar product*. In the first step we simply FOIL, as you may have called it in middle school (and if you didn't, don't worry about it–you know what to do!). We then use the commutativity of the dot product and collect terms to simplify it.

$$(\mathbf{u} + \mathbf{v}) \cdot (\mathbf{u} + \mathbf{v}) = \mathbf{u} \cdot \mathbf{u} + \mathbf{u} \cdot \mathbf{v} + \mathbf{v} \cdot \mathbf{u} + \mathbf{v} \cdot \mathbf{v} = \mathbf{u} \cdot \mathbf{u} + 2(\mathbf{u} \cdot \mathbf{v}) + \mathbf{v} \cdot \mathbf{v} \quad (4.1)$$

It might be tempting to "simplify" $\mathbf{u} \cdot \mathbf{u}$ as "$\mathbf{u}^2$" to make it look more like what happens with real numbers. But this notation suggests that $\mathbf{u}^2$ is some kind of vector, but $\mathbf{u} \cdot \mathbf{u}$ is a scalar, not a vector. In fact $\mathbf{u} \cdot \mathbf{u}$ has a very important meaning that we will now discuss. To see what it is, suppose that $\mathbf{u} = (u_1, u_2)$. Then as in your earlier practice, you see that the distance from the base of $\mathbf{u}$ to its tip may be calculated by applying the Pythagorean Theorem to its components. Visually, that distance is the length of $\mathbf{u}$, which we will denote by $\|\mathbf{u}\|$, and we will alternatively refer to $\|\mathbf{u}\|$ by its more formal algebraic name: the *norm* of $\mathbf{u}$. The Pythagorean Theorem tells us that $\|\mathbf{u}\| = \sqrt{u_1^2 + u_2^2}$. On the other hand $\mathbf{u} \cdot \mathbf{u} = (u_1 u_1 + u_2 u_2) = u_1^2 + u_2^2$. This gives us the following important formula:

$$\mathbf{u} \cdot \mathbf{u} = \|\mathbf{u}\|^2$$

We can now rewrite and simplify Equation (4.1):

$$\|\mathbf{u} + \mathbf{v}\|^2 = \|\mathbf{u}\|^2 + 2\,(\mathbf{u} \cdot \mathbf{v}) + \|\mathbf{v}\|^2$$

Exercise 4.2 *Show in full detail (that is, don't skip steps, including showing where you use the commutative law for the dot product!) the companion formula:*

$$\|\mathbf{u} - \mathbf{v}\|^2 = \|\mathbf{u}\|^2 - 2\,(\mathbf{u} \cdot \mathbf{v}) + \|\mathbf{v}\|^2 \tag{4.2}$$

Your argument should be single a long string of equations starting with:

$$\|\mathbf{u} - \mathbf{v}\|^2 = (\mathbf{u} - \mathbf{v}) \cdot (\mathbf{u} - \mathbf{v})$$

$$= \mathbf{u} \cdot \mathbf{u} + (-\mathbf{v}) \cdot (-\mathbf{v}) + \mathbf{u} \cdot (-)\mathbf{v} + (-\mathbf{v}) \cdot \mathbf{u} + (-\mathbf{v}) \cdot (-\mathbf{v}) = \cdots \tag{4.3}$$

and ending with $\cdots = \|\mathbf{u}\|^2 - 2\,(\mathbf{u} \cdot \mathbf{v}) + \|\mathbf{v}\|^2$. *You should **not** do what I will call a "parallel argument" as you probably learned in high school. That is, do **not** write*

$$\|\mathbf{u} - \mathbf{v}\|^2 = (\mathbf{u} - \mathbf{v}) \cdot (\mathbf{u} - \mathbf{v}) \tag{4.4}$$

$$\|\mathbf{u} - \mathbf{v}\|^2 = \mathbf{u} \cdot \mathbf{u} + (-\mathbf{v}) \cdot (-\mathbf{v}) + \mathbf{u} \cdot (-)\mathbf{v} + (-\mathbf{v}) \cdot \mathbf{u} + (-\mathbf{v}) \cdot (-\mathbf{v})$$

$$\vdots$$

in which you manipulate the sides of an equation. While this may have worked fine in high school, this approach can lead to logical errors in more advanced computations. I will not elaborate on this now–just try to develop this better, safer habit. If you are not sure about this, go ahead and write down a "parallel argument" in which the left side of the equation stays the same from top to bottom. Then just erase all the "left sides" except for the first one. For example, you may get the start of (4.3) by erasing the second $\|\mathbf{u} - \mathbf{v}\|^2$ *in (4.4).*

Exercise 4.3 *For vectors $\mathbf{u}$ and $\mathbf{v}$ in the plane, show that*

1. $\|\mathbf{u} + \mathbf{v}\|^2 + \|\mathbf{u} - \mathbf{v}\|^2 = 2(\|\mathbf{u}\|^2 + \|\mathbf{v}\|^2)$ (Again, you should not do a "parallel argument", but write down a long string of equalities.)

2. Show that the sum of the squares of the lengths of the diagonals of a parallelogram is equal to the sum of the squares of the lengths of its sides. This is known as the Parallelogram Law. Hint: How can the diagonals be geometrically represented using $\mathbf{u} - \mathbf{v}$ and $\mathbf{u} + \mathbf{v}$?

The computations from the above exercises have a very important consequence. Given vectors $\mathbf{u}$ and $\mathbf{v}$, we can form a triangle with $\mathbf{u}$, $\mathbf{v}$ and $\mathbf{u}-\mathbf{v}$. On the one hand, since the lengths of the three sides are $\|\mathbf{u}-\mathbf{v}\|$, $\|\mathbf{u}\|$, and $\|\mathbf{v}\|$, the Law of Cosines (which may be viewed as the version of the Pythagorean Theorem that works for triangles that aren't right triangles) looks like the following, where θ is the angle between the vectors $\mathbf{u}$ and $\mathbf{v}$:

$$\|\mathbf{u}-\mathbf{v}\|^2 = \|\mathbf{u}\|^2 + \|\mathbf{v}\|^2 - 2\|\mathbf{u}\|\|\mathbf{v}\|\cos\theta$$

(If you haven't seen it in awhile, look up the Law of Cosines online and see how it is proved geometrically using the definition of the cosine function. Also note that when $\theta = \frac{\pi}{2}$, i.e. the triangle is a right triangle, $\cos\theta = 0$ and you get precisely the Pythagorean Theorem.) Comparing this expression with Equation (4.2), we obtain another important formula:

$$\mathbf{u}\cdot\mathbf{v} = \|\mathbf{u}\|\|\mathbf{v}\|\cos\theta \tag{4.5}$$

where θ is the angle between the vectors $\mathbf{u}$ and $\mathbf{v}$. This means that we have two different ways to understand and compute the dot product. The "algebraic" way, which was our original definition and only involves multiplying and adding components, and the formula above, which has a highly geometric interpretation. In particular, if we have plane vectors $\mathbf{u}$ and $\mathbf{v}$ and would like to find the angle between them, we may use

$$\cos\theta = \frac{\mathbf{u}\cdot\mathbf{v}}{\|\mathbf{u}\|\|\mathbf{v}\|} \tag{4.6}$$

Of course this last formula doesn't make sense if either $\|\mathbf{u}\| = 0$ or $\|\mathbf{v}\| = 0$, but in this case either $\mathbf{u}$ or $\mathbf{v}$ is the 0-vector, so the "angle between" it and another vector also doesn't make geometric sense. In keeping with the theme of not computing ugly decimal numbers, you will never be asked in MPs to actually compute θ, unless it has a simple value as when θ is a multiple of π by an integer or "nice fraction" like $\pm\frac{1}{2}$, $\pm\frac{1}{4}$, or $\pm\frac{1}{3}$. If you really need a decimal approximation of θ, you can find it using the $\cos^{-1}$ button on your calculator or computational software. You probably don't need any more practice pushing the $\cos^{-1}$ button!

Formula (4.5) and the definition of the dot product together make an important connection between algebra and geometry, and you should certainly memorize them both. Better yet, use them so much that it will be hard to forget them!

4.2   Vectors and Norms in Euclidean Space

The 0-vector in $\mathbb{R}^n$ is the vector $\mathbf{0} = (0,\ldots,0)$. If $\mathbf{v} = (v_1,\ldots,v_n)$ and $\mathbf{w} = (w_1,\ldots,w_n)$ are vectors in $\mathbb{R}^n$, then we define

$$\mathbf{v}+\mathbf{w} = (v_1+w_1,\ldots,v_n+w_n)$$

That is, the components of the sum are the sums of the components. This is the same definition of addition that we would get considering $\mathbf{v}$ and $\mathbf{w}$ as row (or column) vectors, so the same properties hold as for matrices, as you saw for vectors in the plane. Similarly, if k is a scalar, then $k\mathbf{v} = (kv_1, \ldots, kv_2)$ is the scalar multiple of $\mathbf{v}$ by k. Likewise, the *dot product* (also known as the *scalar product*) is defined by $\mathbf{v} \cdot \mathbf{w} = v_1 w_1 + v_2 w_2 + \cdots + v_n w_n$. This expression of the dot product will be referred to as the "coordinate expression". Very often, when doing calculations involving the dot product, the coordinate expression is *not* the way to proceed. I will occasionally warn you not to try such an approach, but rather to use the "coordinate free" properties of the dot product that we will see below: positive definiteness, symmetry, and bilinearity.

As in the plane, with essentially the same proof, the dot product in $\mathbb{R}^n$ is commutative. It is convenient to use summation notation: $\mathbf{v} \cdot \mathbf{w} = \sum_{i=1}^{n} v_i w_i$.

MP 47 *DOT PRODUCT*

Next we define $\|\mathbf{v}\| = \sqrt{v_1^2 + \cdots + v_n^2} = \sqrt{\sum_{i=1}^{n} v_i^2}$, and refer to $\|\mathbf{v}\|$ as the *norm* of $\mathbf{v}$. By definition, $\|\mathbf{v}\| = \sqrt{\mathbf{v} \cdot \mathbf{v}}$ As in the plane, we also refer to $\|\mathbf{v}\|$ as the *length* of $\mathbf{v}$. We will justify the latter term below. It is often more convenient to avoid the square root until absolutely necessary, and work instead with $\|\mathbf{v}\|^2 = \sum_{i=1}^{n} v_i^2$. You saw this already, for example, in Exercise 4.2.

MP 48 *NORM*

The same argument as in the plane shows that $\|k\mathbf{v}\| = |k| \|\mathbf{v}\|$. Since the sum of non-negative numbers is non-negative, $\|\mathbf{v}\| \geq 0$ for any $\mathbf{v}$. Of course if $\mathbf{v} = \mathbf{0} = (0, 0, \ldots, 0)$ then by direct calculation, $\|\mathbf{v}\| = \sqrt{v_1^2 + \cdots + v_n^2} = \sqrt{0^2 + \cdots + 0^2} = 0$. On the other hand, suppose that $\|\mathbf{v}\| = 0$. This means that

$$0 = \|\mathbf{v}\|^2 = \sum_{i=1}^{n} v_i^2$$

If any $v_i^2 > 0$ for any v_i then since all of the numbers v_i^2 are non-negative, it would be impossible for $\sum_{i=1}^{n} v_i^2 = 0$ (because there would have to be a negative term to "cancel" the positive term v_i^2). This proves the following important fact:

Theorem 4.1 *The norm function is positive definite. That is, for any vector $\mathbf{v}$, $\|\mathbf{v}\| \geq 0$, and $\|\mathbf{v}\| = 0$ if and only if $\mathbf{v} = \mathbf{0}$.*

Now consider what the *angle* between two non-zero vectors should mean. Visualize two vectors $\mathbf{v}$ and $\mathbf{w}$ based at the same point sitting in $\mathbb{R}^3$, at first not pointing in the same direction or in opposite directions. So $\mathbf{v}$ and $\mathbf{w}$ are visualized as two arrows from the same base point, pointing in directions that do not align with one another. It shouldn't be hard to see that $\mathbf{v}$ and $\mathbf{w}$ "span" a standard parallelogram from plane geometry, in a sense that we will make precise later. For example, your computer screen is a good approximation of a piece of a plane sitting in space, and you could, in principle, shift and rotate your computer so that those two vectors "lie in" the flat screen. Having done this, you could then measure the angle between those two vectors as you learned to do in the plane. That is,

$$\cos\theta = \frac{\mathbf{v}\cdot\mathbf{w}}{\|\mathbf{v}\|\,\|\mathbf{w}\|} \tag{4.7}$$

What we are doing now is using this formula, which you saw for $\mathbb{R}^2$, to *define* the angle between any two non-zero vectors in any Euclidean space. Your geometric intuition should tell you that this is the "right way" to define the angle θ. As mentioned previously, we aren't "solving for" θ by taking the inverse cosine of each side. Equation (4.7) defines θ *implicitly*, and as you may recall from calculus, many times it is better to work with an implicit formula for a variable than to solve for that variable the beginning. For example, in *implicit differentiation* from your calculus class, you take the derivative *before* you "solve for" the dependent variable, which can be very useful in some situations.

The other reason to not "solve for θ" in the above equations is that $\cos^{-1}\theta$ can take infinitely values for any θ, depending on the context. Your calculator by default returns only values of θ in radians $0 \le \theta \le \pi$. Many situations in physics and engineering involve periodic values, and in such situations, you might need a value of θ in some other interval, such as $\pi \le \theta \le 2\pi$. That is, your calculator will return a decimal approximation of $\frac{\pi}{2}$ for $\cos^{-1}(0)$, but maybe your physical application requires $\frac{3\pi}{2}$, which also satisfies $\cos\frac{3\pi}{2} = 0$. It is generally simpler and easier to just leave the definition of θ implicit as in Equation (4.7) and sort things out later for a specific application.

To be clear, in this course the angle θ between vectors is always measured in *radians*, and $0 \le \theta \le \pi$. Yes, many people, including engineers and other scientific professionals, still use degrees! The problem with degrees is that calculus gets messy—for example, it is *not* true that if $f(\theta) = \sin\theta$ then $f'(\theta) = \cos\theta$, if the angle is measured in degrees. So if you insist on using degrees but you want to use the formulas from calculus that you learned, then the simplest thing to do is do your calculus with radians and then convert to degrees at the end. For basic Euclidean geometry without calculus, degrees *are* convenient because the people who invented these units in antiquity had the foresight to choose 360 degrees for the full sweep the circle—and 360 is wholly divisible by 2, 3, 4, 5, 6, 8, 9, 10, and 12, to name a few divisors. So when it comes to basic geometry, degrees are definitely convenient, but when it comes to calculus, radians are the best units to use.

MP 49 *ANGLE*

When $\|\mathbf{v}\| = 0$ or $\|\mathbf{w}\| = 0$, on the one hand Equation (4.7) doesn't make sense, and on the other hand, by Theorem 4.1, this also means that $\mathbf{v}$ or $\mathbf{w}$ is the 0-vector, which doesn't "point" in any direction, so its angle with any other vector doesn't make geometric sense–as in the plane.

Speaking of directions, we might as well formally define what they are. If $\mathbf{v}$ is a non-zero vector, then $\|\mathbf{v}\| \neq 0$ and we can consider the vector $\frac{1}{\|\mathbf{v}\|}\mathbf{v}$. That is, we are multiplying the vector $\mathbf{v}$ by the scalar that is the reciprocal of its length. This notation is often simplified by writing $\frac{\mathbf{v}}{\|\mathbf{v}\|}$. Since $\|\mathbf{v}\|$ is always non-negative, $\|\|\mathbf{v}\|\| = \|\mathbf{v}\|$, and therefore if $\|\mathbf{v}\| \neq 0$, $\left|\frac{1}{\|\mathbf{v}\|}\right| = \frac{1}{\|\mathbf{v}\|}$. Now let's calculate

$$\left\|\frac{\mathbf{v}}{\|\mathbf{v}\|}\right\| = \left\|\frac{1}{\|\mathbf{v}\|}\mathbf{v}\right\| = \left|\frac{1}{\|\mathbf{v}\|}\right|\|\mathbf{v}\| = \frac{1}{\|\mathbf{v}\|}\|\mathbf{v}\| = 1$$

Geometrically speaking, this calculation shows that if you take a non-zero vector and "divide by its length", you get a vector of length 1. A vector $\mathbf{u}$ such that $\|\mathbf{u}\| = 1$ is called a *unit vector*. If $\mathbf{v} \neq \mathbf{0}$, then $\frac{\mathbf{v}}{\|\mathbf{v}\|}$ is called the *unit vector in the direction of* $\mathbf{v}$, or more simply, the *direction of* $\mathbf{v}$. To reiterate, the 0-vector has no direction–both geometrically and in terms of Equation (4.7).

There is a subtle but important point that needs to be addressed. Since the right side of Equation (4.7) is supposed to be the cosine of an angle, which takes values between -1 and 1, in order for Equation (4.7) to make sense, the right hand side of the equation *must* always satisfy.

$$-1 \leq \frac{\mathbf{v}\cdot\mathbf{w}}{\|\mathbf{v}\|\,\|\mathbf{w}\|} \leq 1 \tag{4.8}$$

Put another way, if you try to enter a number not between -1 and 1 in your calculator and press the $\cos^{-1}$ button, it will return an error. So in order to use Equation (4.7) to define the angle in higher dimensions, we *must* prove that the inequality (4.8) is always true in $\mathbb{R}^n$. By basic algebra, this inequality, as long as $\mathbf{v}$ and $\mathbf{w}$ are non-zero vectors, is equivalent to

$$-\|\mathbf{v}\|\,\|\mathbf{w}\| \leq \mathbf{v}\cdot\mathbf{w} \leq \|\mathbf{v}\|\,\|\mathbf{w}\|$$

which in turn means

$$|\mathbf{v}\cdot\mathbf{w}| \leq \|\mathbf{v}\|\,\|\mathbf{w}\| \tag{4.9}$$

This is a very important inequality called the *Cauchy-Schwarz Inequality*. If you have not worked much with inequalities and absolute values, you should take a moment to check for yourself that the Cauchy-Schwarz Inequality can be equivalently written as

$$-\|\mathbf{v}\|\,\|\mathbf{w}\| \leq \mathbf{v}\cdot\mathbf{w} \leq \|\mathbf{v}\|\,\|\mathbf{w}\|$$

This is a special case of the general fact that you probably saw in calculus but can check for yourself using cases: for any real number x and non-negative real number M, $|x| \leq M$ is equivalent to $-M \leq x \leq M$. The other important point about the Cauchy-Schwarz Inequality is that, unlike Inequality (4.8), it is perfectly fine for $\mathbf{v}$ or $\mathbf{w}$ to be the zero vector in the Cauchy-Schwarz inequality.

To reiterate, without the Cauchy-Schwarz Inequality, we have no guarantee that our definition of angle between non-zero vectors in Equation 4.7) *even makes sense!* If you decided to try to prove this important inequality, you might be tempted to start with the Cauchy-Schwarz equation as expressed as a function of the vector components:

$$\left| \sum_{i=1}^{n} v_i w_i \right| \leq \sqrt{\sum_{i=1}^{n} w_i^2} \sqrt{\sum_{i=1}^{n} v_i^2} \tag{4.10}$$

If this looks ugly and difficult, that suggests that this might not be the best way to prove it. Others who have worked out the proof long ago have found a simpler approach. But one core idea is important: use a "component free" approach that utilizes the properties of the dot product, not the formula for the dot product. We will also start with the special case when $\|\mathbf{v}\| = \|\mathbf{w}\| = 1$. This is a common step in calculations involving vectors sometimes referred to as "normalization". In this case the Cauchy-Schwarz Inequality already looks a lot simpler:

$$|\mathbf{v} \cdot \mathbf{w}| \leq 1 \tag{4.11}$$

As discussed above this is equivalent to

$$-1 \leq \mathbf{v} \cdot \mathbf{w} \leq 1 \tag{4.12}$$

To prove this vastly simpler inequality, we start with something that we know is non-negative by Theorem 4.1, namely the dot product of any vector with itself. In this case we use $\mathbf{v} - \mathbf{w}$:

$$0 \leq (\mathbf{v} - \mathbf{w}) \cdot (\mathbf{v} - \mathbf{w}) = \|\mathbf{v}\|^2 - 2(\mathbf{v} \cdot \mathbf{w}) + \|\mathbf{w}\|^2 = 1 - 2(\mathbf{v} \cdot \mathbf{w}) + 1$$

This simplifies to $2\,(\mathbf{v} \cdot \mathbf{w}) \leq 2$ or $\mathbf{v} \cdot \mathbf{w} \leq 1$. This is the right-hand inequality in (4.12). The other inequality can be obtained by basically the same argument and is an exercise below. This completes the proof of the Cauchy-Schwarz Inequality for unit vectors. Now suppose that $\mathbf{v}$ and $\mathbf{w}$ are any non-zero vectors. Then $\frac{\mathbf{v}}{\|\mathbf{v}\|}$ and $\frac{\mathbf{w}}{\|\mathbf{w}\|}$ are unit vectors, so we can apply to them the Inequality (4.11) that we just proved:

$$\left| \frac{\mathbf{v}}{\|\mathbf{v}\|} \cdot \frac{\mathbf{w}}{\|\mathbf{w}\|} \right| \leq 1$$

By the properties of the dot product we can "pull out" the scalars $\frac{1}{\|\mathbf{v}\|}$ and $\frac{1}{\|\mathbf{w}\|}$ to get

$$\frac{1}{\|\mathbf{v}\|}\frac{1}{\|\mathbf{w}\|}\,|\mathbf{v}\cdot\mathbf{w}| \le 1 \Leftrightarrow |\mathbf{v}\cdot\mathbf{w}| \le \|\mathbf{v}\|\,\|\mathbf{w}\|$$

There is one more case to check (were you keeping track?). What if either $\|\mathbf{v}\| = 0$ or $\|\mathbf{w}\| = 0$? In that case we can just directly check the Cauchy-Schwarz inequality, since we also know that in this case $\mathbf{v}\cdot\mathbf{w} = 0$ and the Cauchy-Schwarz Inequality reduces to $0 \le 0$, which is always true.

Exercise 4.4 *Finish the proof of the Cauchy-Schwarz Inequality by starting with $0 \le (\mathbf{v}+\mathbf{w})\cdot(\mathbf{v}+\mathbf{w})$ using that to obtain $-1 \le \mathbf{v}\cdot\mathbf{w}$.*

If the proof of the Cauchy-Schwarz Inequality seems to you to involve some trickery that you would not have thought of on your own, keep in mind that experienced mathematicians had centuries to figure these things out. These steps that you may regard as "tricks" are better called "techniques" that have developed over centuries, and are useful in many contexts. However, there is one definite takeaway that is broadly relevant to this course: using "coordinate-free" computations can be simpler.

Remark 4.1 *There is another important point to be made. Because we avoided using the component expression (4.10) for the Cauchy-Schwarz Inequality and only used the general algebraic properties of the dot product, the exact same proof will work with any function that has the properties of the dot product, regardless of how it is actually computed. So the proof that you just worked through works in any space in which there is some kind of scalar product, including infinite dimensional spaces of functions in which "dot products" are computed using integration from calculus.*

An important property of the dot product that we have used is *bilinearity*. For any vectors $\mathbf{u},\mathbf{v},\mathbf{w}$ and scalar k,

$$(\mathbf{u}+k\mathbf{v})\cdot\mathbf{w} = \mathbf{u}\cdot\mathbf{w} + k\,(\mathbf{u}\cdot\mathbf{v}) \tag{4.13}$$

You can easily check this directly, which is bilinearity in the "first slot". We have already checked that the dot product is commutative, and from that property we can now derive bilinearity in the "second slot":

$$\mathbf{u}\cdot(\mathbf{w}+k\mathbf{v}) = \mathbf{u}\cdot\mathbf{w} + k\,(\mathbf{v}\cdot\mathbf{u})$$

Exercise 4.5 *Use commutativity of the dot product and Equation (4.13) to show "bilinearity in the second slot". You could of course do the computation directly, but that's not the point of this exercise. You should start with $\mathbf{u}\cdot(\mathbf{v}+k\mathbf{w})$, use commutativity of the dot product to "change slots", apply Equation 4.13, then use commutativity to "change slots" again. You should be able to do this with one long string of equations, justifying each "=" as needed.*

Commutativity of the dot product is more commonly referred to as "symmetry". The final defining factor of dot products is the positive definiteness of Theorem 4.1, but expressed in terms of the dot product, not the norm. That is for any vector $\mathbf{v}$, $\mathbf{v} \cdot \mathbf{v} \geq 0$, and $\mathbf{v} \cdot \mathbf{v} = 0$ if and only if $\mathbf{v} = \mathbf{0}$. We will now collect all of these properties into a theorem. You should review the proof of the Cauchy-Schwarz Inequality to see that it only depends on these properties, and not the component definition of the dot product.

Theorem 4.2 *The dot product satisfies the following properties, for all vectors* $\mathbf{u}$, $\mathbf{v}$, *and* $\mathbf{w}$ *and scalars* k:

      1. *(Positive Definiteness)* $\mathbf{v} \cdot \mathbf{v} \geq 0$, *and* $\mathbf{v} \cdot \mathbf{v} = 0$ *if and only if* $\mathbf{v} = \mathbf{0}$

      2. *(Symmetry)* $\mathbf{v} \cdot \mathbf{w} = \mathbf{w} \cdot \mathbf{v}$

      3. *(Bilinearity)* $(\mathbf{u} + k\mathbf{v}) \cdot \mathbf{w} = \mathbf{u} \cdot \mathbf{w} + k(\mathbf{v} \cdot \mathbf{w})$

There is one more very important theorem in this section.

Theorem 4.3 *(The Triangle Inequality) If* $\mathbf{u}, \mathbf{v}$ *are vectors, then*

$$\|\mathbf{u} + \mathbf{v}\| \leq \|\mathbf{u}\| + \|\mathbf{v}\|$$

While proving this theorem, you should check each step in the proof to see that we are only using Positive Definiteness, Symmetry, Bilinearity, and the Cauchy-Schwarz Inequality, which itself was proved only using those three properties of the dot product. What this means is that this proof is valid for any function that has the properties listed in Theorem 4.2. This means that you won't have to do these proofs all over again in the future when you move beyond Euclidean spaces to more general "inner product spaces" or "scalar product spaces".

The proof is a straightforward coordinate-free calculation (which would be a nightmare using the coordinate definition of the dot product). The calculation will be expressed as a long string of equals and inequalities all going in the "same direction" (in this case $\leq$), with justification at the bottom for those steps that are not just basic algebra. Note that we calculate using $\|\mathbf{u} + \mathbf{v}\|^2$ rather than $\|\mathbf{u} + \mathbf{v}\|$ since this allows us to avoid bothering with those square roots!

$$\|\mathbf{u} + \mathbf{v}\|^2 = (\mathbf{u} + \mathbf{v}) \cdot (\mathbf{u} + \mathbf{v}) = \|\mathbf{u}\|^2 + 2(\mathbf{u} \cdot \mathbf{v}) + \|\mathbf{v}\|^2 \tag{4.14}$$

$$\overset{(1)}{\leq} \|\mathbf{u}\|^2 + 2|\mathbf{u} \cdot \mathbf{v}| + \|\mathbf{v}\|^2 \overset{(2)}{\leq} \|\mathbf{u}\|^2 + 2\|\mathbf{u}\|\,\|\mathbf{v}\| + \|\mathbf{v}\|^2 = (\|\mathbf{u}\| + \|\mathbf{v}\|)^2$$

Inequality (1) is the fact that for any real number x, $x \leq |x|$, and Inequality (2) is the Cauchy-Schwarz Inequality. Since both sides are non-negative we can then just take the square root of each side to get the triangle inequality. The formula in the next exercise is known as the "polarization formula", and it is important because it shows that the dot product and the norm essentially determine one another. That is, if you know how to calculate the dot product then you can calculate the norm via the definition $\|\mathbf{v}\|^2 = \mathbf{v} \cdot \mathbf{v}$. Conversely, if you know how to calculate the norm, then you can calculate the dot product using the polarization formula.

Exercise 4.6 *Prove the following equation.*

$$\mathbf{u} \cdot \mathbf{v} = \frac{1}{4} \left(\|\mathbf{u} + \mathbf{v}\|^2 - \|\mathbf{u} - \mathbf{v}\|^2 \right) \tag{4.15}$$

Hint: Again, do not attempt this with a coordinate calculation! Instead, compute $\|\mathbf{u} + \mathbf{v}\|^2$ and $\|\mathbf{u} - \mathbf{v}\|^2$ from the dot product definition and then do some algebra.

For the last topic in this section we need to revisit the idea of the dot product $\mathbf{u} \cdot \mathbf{v}$ as a special case of matrix multiplication. We can define the matrix product $\mathbf{u}\mathbf{v}$ only if $\mathbf{u}$ is a row vector and $\mathbf{v}$ is a column vector, and by definition the 1×1 matrix $\mathbf{u}\mathbf{v}$ is equal to $[\mathbf{u} \cdot \mathbf{v}]$. Suppose that $\mathbf{A}$ is an $n \times n$ matrix. To multiply by $\mathbf{A}\mathbf{u}$ it is necessary for $\mathbf{u}$ to be a column vector, but then we cannot multiply $(\mathbf{A}\mathbf{u})\mathbf{v}$ without taking the transpose of $\mathbf{A}\mathbf{u}$ to make it a row vector. Then we can calculate using the product formula for the transpose:

$$[\mathbf{A}\mathbf{u} \cdot \mathbf{v}] = (\mathbf{A}\mathbf{u})^T \mathbf{v} = \left(\mathbf{u}^T \mathbf{A}^T \right) \mathbf{v} = \mathbf{u}^T \left(\mathbf{A}^T \mathbf{v} \right) = [\mathbf{u} \cdot \mathbf{A}^T \mathbf{v}] \tag{4.16}$$

This means that the one and only entry of the matrix on the left is equal to the entry of the matrix on the right. This gives us the following formula that we will use many times:

$$\mathbf{A}\mathbf{u} \cdot \mathbf{v} = \mathbf{u} \cdot \mathbf{A}^T \mathbf{v} \tag{4.17}$$

Exercise 4.7 *Without redoing the calculation in 4.16, use symmetry of the dot product to derive $\mathbf{u} \cdot \mathbf{A}\mathbf{v} = \mathbf{A}^T \mathbf{u} \cdot \mathbf{v}$ from (4.17).*

4.3   Lines Determined by Points and Vectors

We have been making an effort to distinguish between vectors (which can be added) and points (which cannot). The transition back and forth between these two concepts is facilitated by the following constructions. First, given any two different points $P_1(x_1, \ldots, x_n)$ and $P_2(y_1, \ldots, y_n)$ we will define the vector from P_1 to P_2 as $\overrightarrow{P_1 P_2} = (y_1 - x_1, \ldots, y_n - x_n)$. Put geometrically, $\overrightarrow{P_1 P_2}$ is the vector with the property that when its base is at P_1, its tip is at P_2.

MP 50 *VECTORS BETWEEN POINTS*

In order to properly understand lines, we need to introduce a concept that effectively allows us to consider a point as a vector. To do so, we will introduce the *position vector* of a point $P(x_1, \ldots, x_n)$, which is the vector based at the origin $O = (0, \ldots, 0)$ and tip at the point P. It is easy to calculate the components of the position vector: $\overrightarrow{OP} = (x_1 - 0, \ldots, x_n - 0) = (x_1, \ldots, x_n)$. That is, the position vector $\overrightarrow{OP}$ has exactly the same components as P does coordinates. Why are we introducing what seems to be just another notation for points? This is because we are maintaining the healthy policy that vectors can be added, but points cannot. This distinction between points and vectors becomes more important in more advanced work when the points in question are in some other space than $\mathbb{R}^n$, such as when considering tangent vectors to a surface in $\mathbb{R}^3$, which you will have seen if you had multi-variable (usually third semester) calculus. In those settings, adding points doesn't always make sense, but adding vectors always does. Keeping the distinction between points and vectors now will likely spare you some confusion later.

MP 51 *POSITION VECTORS*

We begin by seeing how to obtain a vector equation for the *line* through two different points P_1 and P_2 in $\mathbb{R}^n$. You can visualize this as follows. The vector $\overrightarrow{OP_1}$ points from the origin O to the point P_1. Place the vector $\mathbf{v} = \overrightarrow{P_1P_2}$ (which is non-zero because $P_1 \neq P_2$) with its base at P_1, so its tip is at P_2. Then the position vector of P_2 satisfies

$$\overrightarrow{OP_2} = \overrightarrow{OP_1} + \mathbf{v}$$

Consider now the vector equation

$$\mathbf{L}(t) = \overrightarrow{OP_1} + t\mathbf{v}$$

where t is a scalar. Then $\mathbf{L}(0) = \overrightarrow{OP_1}$ and $\mathbf{L}(1) = \overrightarrow{OP_1} + \mathbf{v} = \overrightarrow{OP_2}$. Visually speaking, for values of $t > 1$, the vector $\mathbf{L}(t)$ stretches out in the same direction as $\mathbf{v}$ along the line L that visually contains $\mathbf{v}$. For values $0 \leq t \leq 1$, the length of $t\mathbf{v}$ varies from 0 to 1, and $\mathbf{L}(t)$ "moves along" the line segment from P_1 to P_2, and for $t > 1$ $\mathbf{L}(t)$ continues along L. When $t < 0$, $\mathbf{L}(t)$ moves along the line L in the opposite direction away from P_1. The point of this geometric description is not to prove anything, but to intuitively justify why $\mathbf{L}(t)$ really does represent the equation of the line containing P_1 and P_2.

At the same time, there is no need to start with two points. We can carry out the same procedure, with the same justification, if we simply are given a point P and a vector $\mathbf{w}$. That is, define the *line through P in the direction of $\mathbf{w}$* to be $\mathbf{L}(t) = \overrightarrow{OP} + t\mathbf{w}$. Note that $\mathbf{L}(t)$ is always a *position vector* because

$\overrightarrow{OP} + t\mathbf{w}$ has its base at O. In other words, the vector valued equation $\mathbf{L}(t)$ describes the position of a point in $\mathbb{R}^n$ at "time" t, and that position is somewhere on the line through P in the direction of $\mathbf{v}$. I have put "time" in quotes because in most applications t does not actually represent time.

The equation $\mathbf{L}(t) = \overrightarrow{OP} + t\mathbf{w}$, which assigns to each t a position vector in $\mathbb{R}^n$ may seem novel to you, but if you have had enough calculus you will have already seen such functions in $\mathbb{R}^2$ and $\mathbb{R}^3$. (And if you haven't seen it, don't worry about it!) The difference between what we have just done and what you did in calculus (unless you had a very sophisticated calculus course!) is that your calculus course blurred the distinction between points and their position vectors–something that in this course we are careful not to do. What we do next will reconcile our present description of lines with the usual one from calculus. That is, we will describe the function $\mathbf{L}(t)$ *parametrically*, starting with the special case of $\mathbb{R}^3$. Suppose that $P = (p_1, p_2, p)$ and $\mathbf{v} = (v_1, v_2, v_3)$. Then the components of the position vector $\overrightarrow{OP}$ are the coordinates of the point P; that is, $\overrightarrow{OP} = (p_1, p_2, p_3)$. Moreover, $t\mathbf{v} = (tv_1, tv_2, tv_3)$ and therefore the components of $\mathbf{L}(t)$ are $(p_1 + tv_1, p_2 + tv_2, p_3 + tv_3)$. Since $\mathbf{L}(t)$ is a position vector, its components are the coordinates of its tip. Therefore, the vector-valued function $\mathbf{L}$ is equivalent to the function

$$L(t) = (p_1 + tv_1, p_2 + tv_2, p_3 + tv_3)$$

that takes t to the *point* $(p_1 + tv_1, p_2 + tv_2, p_3 + tv_3)$ in $\mathbb{R}^3$.

Equivalently, we can write three equations, one for each coordinate:

$$x = p_1 + tv_1$$
$$y = p_2 + tv_2$$
$$z = p_3 + tv_3$$

You may recognize such a set of functions from your calculus class, where it may be been referred to as a "parameterized curve" in $\mathbb{R}^3$ (although in this case the "curve" is a straight line!). The individual functions above are called *parametric equations* for the line. As we will see later, these parametric equations are related to the parametric equations you have seen earlier for solutions to homogeneous systems of equations.

Returning to $\mathbb{R}^n$, by the same argument we see that the parameterized equations for a line through a point $P = (p_1, \ldots, p_n)$ in the direction of a non-zero vector $\mathbf{v} = (v_1, \ldots, v_n)$ are

$$x_1 = p_1 + tv_1 \tag{4.18}$$
$$x_2 = p_2 + tv_2$$
$$\vdots$$
$$x_n = p_n + tv_n$$

These equations describe the coordinates of the point moving along the line as a function of t. In coordinate form, we have

$$P(t) = (p_1 + tv_1, p_2 + tv_2, \ldots, p_n + tv_n)$$

Viewed in this way, P is a function from $\mathbb{R}$ into $\mathbb{R}^n$, which we write as P: $\mathbb{R} \to \mathbb{R}^n$.

MP 52 *LINE VIA POINT AND DIRECTION*

One takeaway from the above MPs is that, given a point P and a vector $\mathbf{v}$, the line through P in the direction of $\mathbf{v}$ is the same as the line through P in the direction of $k\mathbf{v}$ for any non-zero scalar k. Intuitively this should make sense—$\mathbf{v}$ and $k\mathbf{v}$ have either the same direction or the opposite direction, and either way they should determine the same line through P. However, the resulting parametric equations will generally not be the same. Let's see how that goes with a specific example. Let $P = (1, 2, 3)$, $\mathbf{v} = (1, -1, 0)$ The parametric equations using $\mathbf{v}$ are

$$x = 1 + t \tag{4.19}$$
$$y = 2 - 2t$$
$$z = 3$$

Note that you can easily read off both P and $\mathbf{v}$ from these equations: The components of $\mathbf{v}$ are the coefficients of the t-terms: $(1, -1, 0)$ and $P = (x(0), y(0), z(0))$. Now let's compute the parametric equations using $k = \frac{1}{3}$, i.e. we take for our vector $\frac{1}{3}\mathbf{v}$. We will use the parameter s to avoid confusion. The new parametric equations are

$$x = 1 + \frac{s}{3} \tag{4.20}$$
$$y = 2 - \frac{s}{3}$$
$$z = 3$$

Again, you can read off the components of $\mathbf{v}$ as the coefficients of the s-term, $(\frac{1}{3}, -\frac{1}{3}, 0)$ and the point P is obtained by setting $s = 0$. We will informally refer to P as the "starting point" of this parameterization. This parameterization is related to the previous one by the linear function $t = \frac{s}{3}$, which is called the *reparameterization function*. You may recall the idea of "change of variables" or "substitution" in integration from calculus, which also ultimately involves reparameterization, but not always with linear functions. The idea here is that both of these equations describe the exact same line, but points on the line are reached at different values of the parameter. For example, the Equations (4.19) at $t = 1$ describe the point $(2, 1, 3)$ and the Equations (4.20) at the value $s = 1$ describe the point $(\frac{4}{3}, \frac{5}{3}, 3)$. The points described are always different except when $s = t = 0$, in which case they both describe the original "starting point" $P(1, 2, 3)$.

As you know, the most general linear function in one variable also has a constant term: $s = kt + a$. What happens if we reparameterize Equations (4.19) with constants $k = 2$ and $a = -1$? To do so we replace t in Equations (4.19) by $s = 2t - 1$. We obtain the equations

$$x = 1 + (2t - 1) = 2t$$
$$y = 2 - (2t - 1) = 3 - 2t$$
$$z = 3$$

If we read off the "starting point" $(t = 0)$ is $P(0, 3, 3)$ for these equations, we see that it is no longer $P(1, 2, 3)$. In fact, the starting point has been shifted precisely by adding $-a\mathbf{v} = (-1, 1, 0)$ to the position vector of P. This illustrates the proof of the following theorem:

Theorem 4.4 *Every line in $\mathbb{R}^n$ has parametric equations of the form*

$$x_1 = p_1 + tv_1 \qquad (4.21)$$

$$\vdots$$

$$x_n = p_n + tv_n$$

where $P(p_1, \ldots, p_n)$ is a point on the line, and $\mathbf{v} = \overrightarrow{PP_1}$ for a point P_1 on the line different from P_1. Any other choice of points P' and P'_1 on the same line results in a linear reparameterization of Equations (4.21). Conversely, any linear reparameterization $s = kt + a$ of Equations (4.21) is of the form

$$x_1 = p'_1 + tv'_1 \qquad (4.22)$$

$$\vdots$$

$$x_n = p'_n + tv'_n$$

where P' is some point on the line and $\mathbf{v}' = \overrightarrow{P'P'_1}$ for some other point P'_1 on the line.

MP 53 *LINE THROUGH TWO POINTS*

4.4 Orthogonality

Vectors $\mathbf{v}$ and $\mathbf{w}$ in $\mathbb{R}^n$ are called *orthogonal* if $\mathbf{v} \cdot \mathbf{w} = 0$. There are two basic situations from a geometric standpoint. First suppose that $\mathbf{v}$ and $\mathbf{w}$

are both non-zero. In that case we can compute the angle θ between them using $\cos\theta = \frac{\mathbf{v}\cdot\mathbf{w}}{\|\mathbf{v}\|\|\mathbf{w}\|} = 0$, which means that $\theta = \frac{\pi}{2}$. That is, $\mathbf{v}$ and are "perpendicular" in the usual sense. If either of $\mathbf{v}$ or $\mathbf{w}$ is $\mathbf{0}$, then it doesn't make sense to discuss the angle between them because one of them has no direction. Therefore it is important to remain mindful that vectors may be orthogonal because one of them is $\mathbf{0}$, and you should not automatically consider them as "perpendicular" unless you know that they are both non-zero. Generally speaking we will use the notation $\mathbf{v} \perp \mathbf{w}$ to mean that $\mathbf{v}$ and $\mathbf{w}$ are orthogonal. This notation is slightly imperfect in the sense that if one of the vectors is $\mathbf{0}$, "$\perp$" is perhaps misleading, but this useful notation and you will soon get used to always considering the possibility that $\mathbf{v} \perp \mathbf{w}$ may mean that one of $\mathbf{v}$ and $\mathbf{w}$ (or both) is the 0-vector.

MP 54 *ORTHOGONALITY*

Two *non-zero* vectors $\mathbf{v}$ and $\mathbf{w}$ in $\mathbb{R}^n$ are said to be *parallel if* $\mathbf{w} = k\mathbf{v}$ for some scalar k. Note that since $\mathbf{v}$ and $\mathbf{w}$ are non-zero, it must be true that $k \neq 0$ and $\mathbf{v} = \frac{1}{k}\mathbf{w}$. The next theorem will lead to a different way to describe lines in the plane. Informally speaking, it says that if two non-zero vectors in the plane are both perpendicular to the same non-zero vector, then they must be parallel.

Theorem 4.5 *If* $\mathbf{u}$ *and* $\mathbf{v}$ *are non-zero vectors in the plane and there is some non-zero vector* $\mathbf{w}$ *such that* $\mathbf{u} \perp \mathbf{w}$ *and* $\mathbf{v} \perp \mathbf{w}$ *then* $\mathbf{u}$ *and* $\mathbf{v}$ *are parallel.*

Proof. The proof is a nice application of the linear algebra you have already learned. Writing $\mathbf{u} = (u_1, u_2)$, $\mathbf{v} = (v_1, v_2)$, and $\mathbf{w} = (w_1, w_2)$ then writing out the equations $\mathbf{u}\cdot\mathbf{w} = 0$ and $\mathbf{v}\cdot\mathbf{w} = 0$ with components, we get the following system of linear equations with w_1 and w_2 as the variables:

$$u_1 w_1 + u_2 w_2 = 0$$
$$v_1 w_1 + v_2 w_2 = 0$$

The matrix corresponding to this system is

$$\mathbf{A} = \begin{bmatrix} u_1 & u_2 \\ v_1 & v_2 \end{bmatrix}$$

Since $\mathbf{w} \neq \mathbf{0}$, $\mathbf{w}$ is a non-trivial solution to this equation. Since the system has a nontrivial solution, according to Theorem 3.10, $0 = \det\mathbf{A} = u_1 v_2 - v_1 u_2$, which means $u_1 v_2 = v_1 u_2$. Therefore,

$$v_2\mathbf{u} = (v_2 u_1, v_2 u_2) = (v_1 u_2, v_2 u_2) = u_2\mathbf{v} \qquad (4.23)$$

As a first case, suppose that v_2 and u_2 are both 0, then since $\mathbf{u}$ and $\mathbf{v}$ are non-zero, v_1 and u_1 are both non-zero. In that case $\mathbf{u} = (u_1, 0) = \frac{u_1}{v_1}\mathbf{v}$, i.e. $\mathbf{u}$

is a non-zero scalar multiple of $\mathbf{v}$. The other case is that one of v_2 or u_2 is not 0. If $v_2 \neq 0$ then we can divide by it in equation (4.23) to obtain $\mathbf{u} = \frac{u_2}{v_2}\mathbf{v}$. The case u_2 is essentially the same–only the letters "u" and "v" need to be swapped. In situations like this, where there are multiple cases that all have basically the same proof, one may use the phrase "without loss of generality" or its acronym WOLOG (pronounced "woe log"). In this proof we would use the phrase like this. "One of v_2 or u_2 is not 0. WOLOG we may assume that $v_2 \neq 0$." ∎

Exercise 4.8 *Go ahead and work out the other case $u_2 \neq 0$ in the above proof to see that it is essentially the same, except for changing letters, as the WOLOG case $v_2 \neq 0$.*

Theorem 4.5 is definitely *not* true in $\mathbb{R}^3$. For example, the unit vectors $\mathbf{e}_1 = (1,0,0)$ and $\mathbf{e}_2 = (0,1,0)$ are not parallel, but are both orthogonal to $\mathbf{e}_3 = (0,0,1)$. But in the plane, Theorem 4.5 has the following consequence:

Theorem 4.6 *Suppose that L_1 and L_2 are lines in the plane through a point P in directions $\mathbf{v}_1$ and $\mathbf{v}_2$, and there is some $\mathbf{w}$ such that $\mathbf{v}_1 \perp \mathbf{w}$ and $\mathbf{v}_2 \perp \mathbf{w}$. Then L_1 and L_2 are the same line.*

Exercise 4.9 *Prove the above theorem (this should only take you one or two sentences).*

We now know the following: Given a point $P(p_1, p_2)$ in the plane and a non-zero vector $\mathbf{w} = (w_1, w_2)$ at P, there is exactly one line L through P that is *orthogonal* to $\mathbf{w}$, in the sense that for every point Q on L, $\overrightarrow{PQ}$ is orthogonal to $\mathbf{w}$. If (x, y) is a point on L, this means that $(x - p_1, y - p_2) \cdot (w_1, w_2) = 0$. Expanding this we obtain $w_1 x + w_2 y = w_1 p_1 + w_2 p_2$. This shows that every line in the plane not only has the parametric expressions that you have already learned, but also is uniquely in the form of a linear equation $ax + by = c$, where $a = w_1$, $b = w_2$ and $c = w_1 p_1 + w_2 p_2$. So we have recovered, using orthogonality, the fact that you learned long ago: lines are represented by linear equations. The linear equation can also be used to find a point and an orthogonal vector, also called a *normal vector* to the line by the calculation we just made. We state all this as a theorem:

Theorem 4.7 *Every line L in the plane may be described by a linear equation $ax + by = d$. Equivalently, the vector $\mathbf{n} = (a, b)$ is a normal vector to L at every point on L.*

Let's now see what happens if we try the same strategy in $\mathbb{R}^3$. That is, fix a point $P(p_1, p_2, p_3)$ and a non-zero vector $\mathbf{w} = (w_1, w_2, w_3)$, and consider the set of all points $Q(x, y, z)$ such that $\overrightarrow{PQ} \perp \mathbf{w}$. Doing basically the same calculation in dimension 3, we write $\overrightarrow{PQ} = (x - p_1, y - p_2, z - p_3)$ and

$$0 = (x - p_1, y - p_2, z - p_3) \cdot (w_1, w_2, w_3)$$

Rearranging terms gets us

$$w_1 x + w_2 y + w_3 z = p_1 w_1 + p_2 w_2 + p_3 w_3$$

and we see that as before, the points Q satisfy a linear equation, now with three variables, $ax + by + cz = d$. You can read off the components of a normal vector $(a, b, c) = (w_1, w_2, w_3)$ from this equation, and can find a point P by taking any solution (x, y, z) for this equation. We call the set of these points the *plane through* P and normal to $\mathbf{w} = (a, b, c)$. Soon we will be able to parameterize this plane as we did the line, except with two variables instead of one. Again this should remind you of parameterization of solutions to homogeneous systems.

This connection is made precise by the following calculation. Suppose we have a homogeneous linear system $\mathbf{AX} = \mathbf{0}$. We may write the equation *a la* (2.10) to obtain

$$\begin{bmatrix} a_{11} & a_{12} & \cdots & a_{1n} \\ a_{21} & a_{22} & \cdots & a_{2n} \\ \vdots & \vdots & \vdots & \vdots \\ a_{n1} & a_{n2} & \cdots & a_{nn} \end{bmatrix} \begin{bmatrix} x_1 \\ x_2 \\ \vdots \\ x_n \end{bmatrix} = \begin{bmatrix} a_{11}x_1 + a_{12}x_2 + \cdots + a_{1n}x_n \\ a_{21}x_1 + a_{22}x_2 + \cdots + a_{2n}x_n \\ \vdots \\ a_{n1}x_1 + a_{n2}x_2 + \cdots + a_{nn}x_n \end{bmatrix} = \begin{bmatrix} 0 \\ 0 \\ \vdots \\ 0 \end{bmatrix}$$

Recall that the ith row in the matrix $\mathbf{A}$ was denoted by $\mathbf{A}_i$, which may regarded as a vector. Note that the middle column matrix has as its entries the dot products $\mathbf{A}_i \cdot \mathbf{X}$. That is, this system is equivalent to setting the dot products $\mathbf{A}_i \cdot \mathbf{X}$ equal to 0–meaning that each row is orthogonal to $\mathbf{X}$. We have proved the following theorem:

Theorem 4.8 *If* $\mathbf{AX} = \mathbf{0}$ *is a homogeneous system of linear equations, then* $\mathbf{X}$ *is a solution of the system if and only if* $\mathbf{X}$ *is orthogonal to every row of* $\mathbf{A}$.

MP 55 *ORTHOGONALITY AND SOLUTIONS*

There is a very useful construction known as orthogonal projection of one vector onto another. The usefulness will become more clear later, but for now trust that this is not just an exercise to give you more work! Given two non-zero vectors $\mathbf{u}$ and $\mathbf{v}$ in $\mathbb{R}^n$, we would like to write $\mathbf{u} = t\mathbf{v} + \mathbf{w}$, where $\mathbf{w} \perp \mathbf{v}$. If you imagine that the earth is flat and the sun is infinitely far away (it's easy if you try!) the idea is that the vector $\mathbf{v}$ is lies in the flat ground and the vector $\mathbf{u}$ starts at the same point, pointing upwards in the vertical plane above $\mathbf{u}$ with some angle θ between 0 and π. The sun then makes a shadow vector $\mathbf{s}$ that lies in the line spanned by $\mathbf{u}$, so $\mathbf{s}$ is a scalar multiple of $\mathbf{v}$. If $0 \leq \theta < \frac{\pi}{2}$ then $\mathbf{s}$ points in the same direction as $\mathbf{v}$. When $\theta = \frac{\pi}{2}$ then $\mathbf{v}$ is perpendicular to $\mathbf{u}$ and $\mathbf{s} = \mathbf{0}$, and when $\frac{\pi}{2} < \theta \leq \pi$ then $\mathbf{s}$ points in the opposite direction as $\mathbf{v}$.

In general, from Euclidean geometry we know that the length $\|\mathbf{s}\|$ of $\mathbf{s}$ is equal to $\|\mathbf{u}\| \cos \theta$. On the other hand, we know that $\|\mathbf{u}\| \cos \theta = \frac{\mathbf{u} \cdot \mathbf{v}}{\|\mathbf{v}\|}$. Now we want to write $\mathbf{s}$ as a scalar multiple of $\mathbf{v}$, which has length $\|\mathbf{v}\|$. That is, to obtain $\mathbf{s}$ from $\mathbf{v}$ we need to first take the unit vector $\frac{\mathbf{v}}{\|\mathbf{v}\|}$ and multiply it by the length of $\mathbf{s}$. Putting this all together we see that

$$\mathbf{s} = \left(\frac{\mathbf{u} \cdot \mathbf{v}}{\|\mathbf{v}\|^2}\right) \mathbf{v}$$

We use this geometric description as a reason to make the following definition:

Definition 4.1 *If $\mathbf{u}$ and $\mathbf{v}$ are vectors in $\mathbb{R}^n$, the orthogonal projection of $\mathbf{u}$ along $\mathbf{v}$ is the vector*

$$proj_{\mathbf{v}}(\mathbf{u}) = \left(\frac{\mathbf{u} \cdot \mathbf{v}}{\|\mathbf{v}\|^2}\right) \mathbf{v}$$

Standard linear algebra texts will now ask you to compute a bunch of these. This *Interactive Linear Algebra* will refrain from doing so because, seriously, do you really need more practice putting numbers into a calculator or computer? A much more important strategy from the standpoint of learning (and your future as a STEM professional) is for you to *understand* where this formula comes from. Ideally, you should be able to explain the entire paragraph that precedes the definition, and all the 9th-grade calculation in the world will not help you with that!

Exercise 4.10 *Let $\mathbf{w} := \mathbf{u} - proj_{\mathbf{v}}(\mathbf{u})$. Show that $\mathbf{w} \cdot \mathbf{v} = 0$. Do not attempt to write this out in coordinates! This calculation explains the name "orthogonal projection". We can write $\mathbf{u} = \mathbf{w} + proj_{\mathbf{v}}(\mathbf{u})$, where $\mathbf{w}$ and $\mathbf{v}$ are orthogonal.*

4.5 The Cross Product

The cross product is normally learned in calculus or physics courses, but even if you have seen it before, this section should give you a new perspective on it by relating it to the determinant function. For this section we will use cofactor expansion to compute the determinant, even though we prove that this is correct somewhat later. In fact, this section will serve as a motivation to prove that you can compute the determinant in that way, because cofactor expansion has very nice geometric interpretations. While the definition of the cross product can be extended to higher dimensions, we will confine our attention to the cross product in $\mathbb{R}^3$. Via cofactor expansion, the formula for the cross product is easy to state and remember. First, as is standard in physics and calculus, we will denote the so-called "standard basis" for $\mathbb{R}^3$

by the vectors $\mathbf{i} = (1, 0, 0)$, $\mathbf{j} = (0, 1, 0)$ and $\mathbf{k} = (0, 0, 1)$. You will soon learn what a "basis" is, but you don't need that information now. Now suppose that $\mathbf{u} = (u_1, u_2, u_3)$ and $\mathbf{v} = (v_1, v_2, v_3)$ are vectors. We define the cross product using the following formula

$$\mathbf{u} \times \mathbf{v} = \det \begin{bmatrix} \mathbf{i} & \mathbf{j} & \mathbf{k} \\ u_1 & u_2 & u_3 \\ v_1 & v_2 & v_3 \end{bmatrix} \tag{4.24}$$

If this definition looks OK to you, then you didn't read it carefully enough! After all, we are putting *vectors* in the entries of a matrix (along with numbers). Physicists do this kind of thing sometimes, and it can be useful. The point is that you should just ignore the strangeness and go ahead and formally compute the determinant using cofactor expansion. That is,

$$\mathbf{u} \times \mathbf{v} = \mathbf{i} \det \begin{bmatrix} u_2 & u_3 \\ v_2 & v_3 \end{bmatrix} - \mathbf{j} \det \begin{bmatrix} u_1 & u_3 \\ v_1 & v_3 \end{bmatrix} + \mathbf{k} \det \begin{bmatrix} u_1 & u_2 \\ v_1 & v_2 \end{bmatrix}$$

This makes a bit more sense–it is an alternating sum of scalars times vectors, although the scalars are written on the right. If we correct this notational problem and write out the determinants of the 2x2 matrices, we then get

$$(u_2 v_3 - u_3 v_2)\mathbf{i} - (u_1 v_3 - v_1 u_3)\mathbf{j} + (u_1 v_2 - v_1 u_2)\mathbf{k}$$

Now you see that unlike $\mathbf{u} \cdot \mathbf{v}$, which is a scalar, $\mathbf{u} \times \mathbf{v}$ is a vector (which is sometimes also called the "vector product"). Using the definitions of $\mathbf{i}, \mathbf{j}, \mathbf{k}$ we can now write out the components:

$$\mathbf{u} \times \mathbf{v} = (u_2 v_3 - u_3 v_2, v_1 u_3 - u_1 v_3, u_1 v_2 - v_1 u_2) \tag{4.25}$$

You see that the original Definition (4.24) is much easier to remember, even though strictly speaking it involves what a mathematician would call "abuse of notation"—meaning it isn't quite right, but it is useful even so.

The cross product is quite important in physics, and we will go through some of its properties. These are very easily (but tediously!) verified using the coordinate expression in Equation (4.25).

1. $\mathbf{u} \times \mathbf{v} = -(\mathbf{v} \times \mathbf{u})$ (Skew Symmetry)
2. $\mathbf{u} \times (k\mathbf{v} + \mathbf{w}) = k(\mathbf{u} \times \mathbf{v}) + \mathbf{u} \times \mathbf{w}$
3. $\mathbf{u} \times \mathbf{u} = \mathbf{0}$
4. $\mathbf{u} \times \mathbf{0} = \mathbf{0}$

Exercise 4.11 *Derive the properties 1 and 3 above by formally using properties of the determinant. To see what this means, I'll derive the fourth property as follows. If a matrix has a row of entries then its determinant is 0, and since the cross product is a vector, this must refer to the vector $\mathbf{0}$. This informal derivation does not "prove" these statements, but should help you to* *rapidly derive them when you need them. The informal derivation of second property involves something called multilinearity of the determinant, which we will discuss later.*

There are various interesting equations involving both cross products and dot products. The proofs generally involve simply writing out both sides of the equation using formula (4.25) doing some tedious 9th-grade algebra and comparing sides. There is little to be learned from such proofs, so there's no point in including them here. The simplest of these shows that the cross product of two vectors is orthogonal to either of the two vectors:

$$\mathbf{u} \cdot (\mathbf{u} \times \mathbf{v}) = \mathbf{0}$$

More precisely, the direction of the vector $\mathbf{u} \times \mathbf{v}$ (when $\mathbf{u}$ and $\mathbf{v}$ are non-zero) is obtained using the "right hand rule"–if you imagine unscrewing the lid of a jar as rotating from $\mathbf{u}$ toward $\mathbf{v}$, then $\mathbf{u} \times \mathbf{v}$ points in the "up" direction. This means that $\mathbf{u} \times \mathbf{v}$ is completely determined by its length, and its length is given by the Lagrange's Identity:

$$\|\mathbf{u} \times \mathbf{v}\|^2 = \|\mathbf{u}\|^2 \|\mathbf{v}\|^2 - (\mathbf{u} \cdot \mathbf{v})^2$$

Lagrange's identity together with the right-hand rule gives a complete geometric description of the cross product in terms of its direction and length.

One also has the Triple Vector Product Equation:

$$\mathbf{u} \times (\mathbf{v} \times \mathbf{w}) = (\mathbf{u} \cdot \mathbf{w})\mathbf{v} - (\mathbf{u} \cdot \mathbf{v})\mathbf{w}$$

Next there is the Scalar Triple Product Formula

$$\mathbf{u} \cdot (\mathbf{v} \times \mathbf{w}) = \det \begin{bmatrix} u_1 & u_2 & u_3 \\ v_1 & v_2 & v_3 \\ w_1 & w_2 & w_3 \end{bmatrix} \tag{4.26}$$

There is no "abuse of notation" in this last formula–all the entries of the matrix are numbers, the determinant is a number, and the left side, which is a dot product of two vectors, is a number. Finally, we use Lagrange's Identity and some basic trig identities to obtain a geometric interpretation of the cross product. In fact,

$$\|\mathbf{u} \times \mathbf{v}\|^2 = \|\mathbf{u}\|^2 \|\mathbf{v}\|^2 - (\mathbf{u} \cdot \mathbf{v})^2 = \|\mathbf{u}\|^2 \|\mathbf{v}\|^2 - \|\mathbf{u}\|^2 \|\mathbf{v}\|^2 \cos^2 \theta$$

$$= \|\mathbf{u}\|^2 \|\mathbf{v}\|^2 (1 - \cos^2 \theta) = \|\mathbf{u}\|^2 \|\mathbf{v}\|^2 \sin^2 \theta$$

Taking square roots we get:

$$\|\mathbf{u} \times \mathbf{v}\| = \|\mathbf{u}\| \|\mathbf{v}\| |\sin \theta| \tag{4.27}$$

As you may recall from elementary geometry, the right side is just the area of the parallelogram spanned by $\mathbf{u}$ and $\mathbf{v}$, something we will revisit later.

4.6   Matrices as Linear Transformations

We will need the concept of the *standard basis* for $\mathbb{R}^n$. As mentioned previously, the standard basis in $\mathbb{R}^3$ is often denoted by { $\mathbf{i} = (1,0,0)$, $\mathbf{j} = (0,1,0)$ and $\mathbf{k} = (0,0,1)$}. Since there aren't enough letters in the alphabet to continue this very long, in $\mathbb{R}^n$, the standard basis is

$$\mathbf{e}_1 = (1,0,0,\ldots,0)$$
$$\mathbf{e}_2 = (0,1,0,\ldots,0)$$
$$\vdots$$
$$\mathbf{e}_n = (0,0,0,\ldots,1)$$

Put another way, the standard basis consists of the rows (or columns!) of the identity matrix $\mathbf{I}$. We will later learn what a "basis" is and why $\{\mathbf{e}_1,\ldots,\mathbf{e}_n\}$ is the "standard" one. But for now we note that given a vector $\mathbf{v} = (v_1,\ldots,v_n)$ in $\mathbb{R}^n$, we may write

$$\mathbf{v} = (v_1,0,\ldots,0) + (0,v_2,0,\ldots 0) + \cdots + (0,\ldots,0,v_n)$$

$$= v_1(1,0,\ldots,0) + v_2(0,1,0,\ldots 0) + \cdots + v_n(0,\ldots,0,1)$$

$$= v_1\mathbf{e}_1 + \cdots + v_n\mathbf{e}_n = \sum_{i=1}^{n} v_i\mathbf{e}_i \tag{4.28}$$

We also need a simple theorem that will be used many times. Recall that the jth column of a matrix $\mathbf{A}$ is denoted $\mathbf{A}^j$. This theorem says that you can "extract" the jth column of $\mathbf{A}$ by simply taking the matrix product $\mathbf{Ae}_j$.

Theorem 4.9 *If $\mathbf{A}$ is an $m \times n$ matrix then $\mathbf{Ae}_j = \mathbf{A}^j$, where $\mathbf{e}_j$ is in the standard basis of $\mathbb{R}^n$.*

Exercise 4.12 *Prove the above theorem. This is really just an exercise in writing out the notation for $\mathbf{A}^j$.*

There is a matrix construction that we will use over and over. Given a set $S = \{\mathbf{v}_1,\ldots,\mathbf{v}_m\}$ of vectors, the matrix $\mathbf{C}_S$ consists of the matrix having the vectors $\mathbf{v}_i$ as columns. Using the notation of augmented matrices,

$$\mathbf{C}_S = \begin{bmatrix} \mathbf{v}_1 & | & \mathbf{v}_2 & | & \cdots & | & \mathbf{v}_m \end{bmatrix}$$

For example, as we have already observed, if $S = \{\mathbf{e}_1,\ldots,\mathbf{e}_m\}$ then

$$\mathbf{I} = \begin{bmatrix} \mathbf{e}_1 & | & \mathbf{e}_2 & | & \cdots & | & \mathbf{e}_m \end{bmatrix} = \mathbf{C}_S$$

Given an $m \times n$ matrix $\mathbf{A} = [a_{ij}]$ and a column vector $\mathbf{X} = \begin{bmatrix} x_1 \\ x_2 \\ \vdots \\ x_n \end{bmatrix}$ we can

compute the product $\mathbf{AX} = \mathbf{Y}$ where $\mathbf{Y} = \begin{bmatrix} y_1 \\ y_2 \\ \vdots \\ y_m \end{bmatrix}$. We can consider $\mathbf{X}$ as a

vector in $\mathbb{R}^n$ and $\mathbf{Y}$ as a vector in $\mathbb{R}^m$. In other words, there is yet another interpretation of the matrix $\mathbf{A}$, namely as a *function* from $\mathbb{R}^n$ to $\mathbb{R}^m$.

We need some terminology about functions. If D and C are sets, then by definition a function $f : D \to C$ assigns to every element d of D a uniquely determined element $f(d)$ of C. The set D is called the *domain* of f and the set C is called the *codomain* of f. The set of all possible values $f(d)$ is called the *range*, or *image* of f. With these definitions we see that the $m \times n$ matrix $\mathbf{A}$ can be considered as a function that takes n-vectors to m-vectors. That is, via matrix multiplication, $\mathbf{A}$ corresponds to a function $T_{\mathbf{A}} : \mathbb{R}^n \to \mathbb{R}^m$, which we will call the *linear transformation corresponding to* $\mathbf{A}$. It is extremely important to distinguish between the matrix $\mathbf{A}$ (which is an array of numbers) and the transformation $T_{\mathbf{A}}$, which is a function. As we will see later, the same linear transformation may be represented by different matrices via something called "basis change", and if you do not get into the habit of distinguishing matrices from the transformations that they represent, this later topic may be quite confusing.

Notation 2 *It is important to remain "notation flexible" and write vectors in different ways depending on the context. We will use capital letters to represent the column form of vectors with the same components. For example, if $\mathbf{v} = (v_1, \ldots, v_n)$ and $\mathbf{w} = (w_1, \ldots, w_m)$, we can write $\mathbf{T_A}(\mathbf{v}) = \mathbf{w}$ using column notation: $\mathbf{AV} = \mathbf{W}$, which written out looks like*

The linear transformations of some matrices have special properties. The simplest of all is $\mathbf{A} = \mathbf{0}$, which satisfies $T_{\mathbf{A}}(\mathbf{v}) = \mathbf{0}$ for all $\mathbf{v}$ in $\mathbb{R}^n$. Somewhat graphically speaking, $T_{\mathbf{A}}$ "kills" all vectors. More generally, the set of all vectors $\mathbf{v}$ such that $T_{\mathbf{A}}(\mathbf{v}) = \mathbf{0}$ is called the *kernel* of $T_{\mathbf{A}}$, denoted by $\ker T_{\mathbf{A}}$. Vectors $\mathbf{v}$ in the kernel by definition satisfy the equation

$$\mathbf{A} \begin{bmatrix} v_1 \\ v_2 \\ \vdots \\ v_n \end{bmatrix} = \begin{bmatrix} 0 \\ 0 \\ \vdots \\ 0 \end{bmatrix}$$

and so *they are precisely the solutions to the homogeneous system of equations* $\mathbf{AX} = \mathbf{0}$.

A particularly simple transformation is that of the identity matrix. As you know, for any column vector $\mathbf{X}$, $\mathbf{IX} = \mathbf{XI} = \mathbf{X}$. This means that for any vector $\mathbf{v}$, $T_\mathbf{I}(\mathbf{v}) = \mathbf{v}$. Put another way, $T_\mathbf{I}$ "fixes every vector", or "takes every vector to itself". Scalar multiples of the identity matrix correspond to important transformations known as "dilations". For example, the transformation $T_\mathbf{A}$ for the matrix

$$\mathbf{A} = \begin{bmatrix} 2 & 0 & 0 & 0 \\ 0 & 2 & 0 & 0 \\ 0 & 0 & 2 & 0 \\ 0 & 0 & 0 & 2 \end{bmatrix}$$

takes any vector in $\mathbb{R}^4$ and doubles its length.

Exercise 4.13 *Multiply the above matrix by the vector (x, y, z, w) to verify that it doubles lengths.*

Let's now discuss in more detail the words in the term "linear transformation corresponding to $\mathbf{A}$". The word "transformation" is just another word for "function" that tends to be used for linear functions—but in this context it really it means the same thing. As you saw in Exercise 3.1, functions of the form $f(x) = mx$ are uniquely characterized by their linearity. That is, the functions $f(x) = mx$ are precisely the real functions that satisfy $f(ka+b) = kf(a)+f(b)$. Note that the function $f(x) = mx$ can be considered as multiplication by the 1×1 matrix $[m]$.

Definition 4.2 *A function or transformation $T : \mathbb{R}^n \to \mathbb{R}^m$ is called linear if for any scalar k and vectors $\mathbf{v}$ and $\mathbf{w}$, $T(k\mathbf{v} + \mathbf{w}) = kT(\mathbf{v}) + T(\mathbf{w})$.*

Sometimes the single condition $T(k\mathbf{v} + \mathbf{w}) = kT(\mathbf{v}) + T(\mathbf{w})$ is more easily verified as two conditions: $T(k\mathbf{v}) = kT(\mathbf{v})$ and $T(\mathbf{v}+\mathbf{w}) = T(\mathbf{v})+T(\mathbf{w})$. There is one important fact about linear transformations that should be emphasized:

$$T(\mathbf{0}) = \mathbf{0} \text{ if } T \text{ is any linear transformation}$$

This fact is proved by the following computation:

$$T(\mathbf{0}) = T(\mathbf{0} + \mathbf{0}) = T(\mathbf{0}) + T(\mathbf{0}) = 2T(\mathbf{0})$$

That is, $T(\mathbf{0})$ is equal to twice itself, so all of its components must be twice themselves. That means that all the components of $T(\mathbf{0})$ must be 0.

Theorem 4.10 *If $\mathbf{A}$ is a matrix then $T_\mathbf{A} : \mathbb{R}^n \to \mathbb{R}^m$ is linear, and if $T_\mathbf{A} = T_\mathbf{B}$ then $\mathbf{A} = \mathbf{B}$.*

Proof. For any scalar k and vectors $\mathbf{v}$ and $\mathbf{w}$, we will calculate with column vector form, using the properties of matrix multiplication:

$$T_{\mathbf{A}}(k\mathbf{v} + \mathbf{w}) = \mathbf{A}\left(k\begin{bmatrix} v_1 \\ \vdots \\ v_n \end{bmatrix} + \begin{bmatrix} w_1 \\ \vdots \\ w_n \end{bmatrix}\right)$$

$$k\mathbf{A}\begin{bmatrix} v_1 \\ \vdots \\ v_n \end{bmatrix} + \mathbf{A}\begin{bmatrix} w_1 \\ \vdots \\ w_n \end{bmatrix} = kT_{\mathbf{A}}(\mathbf{v}) + T_{\mathbf{A}}(\mathbf{w})$$

The second part of the theorem is important–it is a kind of uniqueness statement: for any linear transformation T, there can only be one matrix $\mathbf{A}$ such that $T_{\mathbf{A}} = T$. To see why it is true, we need to show that if $\mathbf{Aw} = \mathbf{Bw}$ for every vector $\mathbf{w}$, then $\mathbf{A} = \mathbf{B}$. We can apply this hypothesis using $\mathbf{w} = \mathbf{e}_j$ and compute

$$\mathbf{A}^j = \mathbf{Ae}_j = \mathbf{Be}_j = \mathbf{B}^j.$$

That is, $\mathbf{A}$ and $\mathbf{B}$ have the exact same columns, and so are equal. ■

Now you may be wondering–is the converse of the above theorem true? That is, if I have a linear transformation, is it always of the form $T_{\mathbf{A}}$ for some matrix $\mathbf{A}$? We will see now that the simple answer is "yes", but there are more interesting answers that we will see later. At any rate, we are given a linear transformation $T : \mathbb{R}^n \to \mathbb{R}^m$. How do we find a matrix $\mathbf{A}$ so that $T = T_{\mathbf{A}}$? The way to do this is quite important–you should not ignore this proof because the method actually *tells you how to find the matrix* $\mathbf{A}$, and this process will be used many times in MPs and other parts of this text. Beyond this course, linear transformations show up everywhere in science–and the fact that you can concretely find a matrix that allows you to compute it via matrix multiplication has many practical applications.

Given the linear transformation T, to construct the matrix $\mathbf{A}$, first note that we actually only need to know where T takes the standard basis. In fact, according to the (4.28) we can always write $\mathbf{v} = v_1\mathbf{e}_1 + v_2\mathbf{e}_2 + \cdots + v_n\mathbf{e}_n$. Since T is linear, we have that

$$T(\mathbf{v}) = T(v_1\mathbf{e}_1 + v_2\mathbf{e}_2 + \cdots + v_n\mathbf{e}_n)$$

$$= v_1T(\mathbf{e}_1) + v_2T(\mathbf{e}_2) + \cdots + v_nT(\mathbf{e}_n) = \sum_{i=1}^{n} v_iT(\mathbf{e}_i)$$

It is worth repeating what this means: in order to know the value of $T(\mathbf{v})$ for *any* vector $\mathbf{v}$, we only need to know the n vectors in $S = \{T(\mathbf{e}_1), \ldots, T(\mathbf{e}_n)\}$. Now each $T(\mathbf{e}_i)$ is an m-vector, and there are n of them, so we can create the $m \times n$ matrix $\mathbf{C}_S$, which as you should recall has each $T(\mathbf{e}_i)$ as the ith column.

$$\mathbf{C}_S = \begin{bmatrix} T(\mathbf{e}_1) & | T(\mathbf{e}_2) & | & \cdots & | & T(\mathbf{e}_n) \end{bmatrix}$$

Now let's calculate

$$T_{\mathbf{C}_S}(\mathbf{v}) = \begin{bmatrix} T(\mathbf{e}_1) & | \, T(\mathbf{e}_2) & | & \cdots & | & T(\mathbf{e}_n) \end{bmatrix} \begin{bmatrix} v_1 \\ \vdots \\ v_n \end{bmatrix}$$

$$= v_1 T(\mathbf{e}_1) + v_2 T(\mathbf{e}_2) + \cdots + v_n T(\mathbf{e}_n) = T(\mathbf{v})$$

The above computation may be simplified using "block multiplication" that was mentioned earlier. In this case, the matrix on the left has column vectors for its entries, and the matrix on the right has scalar entries. Since scalars can be multiplied by column vectors, one can do the calculations just as though it were the dot product of a row and column matrix. Another way to think about it is that the scalar v_1 is multiplied by the first entry in each row, which is the same as multiplying the first column $T(\mathbf{e}_1)$ by the scalar v_1. The above calculation proves the following important theorem:

Theorem 4.11 *Let* $T : \mathbb{R}^n \to \mathbb{R}^m$ *be a linear transformation and* $S = \{T(\mathbf{e}_1), \ldots, T(\mathbf{e}_n)\}$. *Then* $T = T_{\mathbf{C}_S}$.

This theorem gives us yet another way to recognize linear functions, since they are all of the form $T_{\mathbf{A}}$ for some matrix $\mathbf{A}$. Writing this out again, we see

$$\mathbf{A} \begin{bmatrix} x_1 \\ \vdots \\ x_n \end{bmatrix} = \begin{bmatrix} a_{11}x_1 + a_{12}x_2 + \cdots + a_{1n}x_n \\ \vdots \\ a_{m1}x_1 + a_{m2}x_2 + \cdots + a_{mn}x_n \end{bmatrix}$$

In other words, each entry of the row vector on the right is itself a linear function of the variables $x_1, \ldots, x_n$. You will practice finding $\mathbf{C}_S$ in the next MP.

MP 56 *MATRIX OF A LINEAR TRANSFORMATION*

If you paid close attention doing the previous MP you might have noticed that you could have just read columns of $\mathbf{C}_S$ off as the coefficients of the linear functions that define L! For example, you were given $T\left(\begin{bmatrix} x_1 \\ x_2 \end{bmatrix} \right) = \begin{bmatrix} 2x_1 - x_2 \\ x_1 \\ x_1 + x_2 \end{bmatrix}$ in one of the problems, and the matrix that you got was

$$\mathbf{C}_S = \begin{bmatrix} 2 & -1 \\ 1 & 0 \\ 1 & 1 \end{bmatrix}$$

So why did I make you do all that work when you could just have read the matrix off from the coefficients of the equations? Because later on we will be using bases other than the standard basis. In that case, the basic *process* that you just practiced still works, but you can't just read off the matrix from the coefficients as you can with the standard basis. The standard basis is nice for some situations, but very often it is not the "best" basis for other purposes. For a proper choice of basis, the matrix corresponding to the transformation can have really nice properties. All of these statements will be made precise later.

4.7   Rotations in the Plane

If $\mathbf{A}$ is a square $n \times n$ matrix, then $T_{\mathbf{A}} : \mathbb{R}^n \to \mathbb{R}^n$ and in many situations it is useful to consider $T_{\mathbf{A}}$ as a transformation of $\mathbb{R}^n$ within itself. To illustrate this idea we will consider rotations in the plane. You already have a geometric understanding of rotations in the plane as being a kind of "rigid motion" of the plane that "turns the plane" around the origin by a particular angle θ, which we will refer to as ρ_θ.

Let's suppose that you were faced with the problem of finding the coordinates of the point $(1, 3)$ in the plane, after a rotation of angle $\frac{\pi}{3}$. Problems of this sort are common in engineering and scientific fields. Using basic Euclidean geometry this would be a bit tricky even though $\pi/3$, which is $60°$, is a relatively "nice" angle and you could try to use a 30-60-90 triangle by bisecting the Euclidean triangle formed. This is a tedious computation that you would have to modify for every different point! Linear algebra offers a much better solution. But let's be absolutely clear what we are doing. We have a geometric idea of "rotation" and we are trying to find a way to describe it mathematically. This has been done long ago, of course, but more complex modern problems have a similar flavor: giving a mathematical description for something that we observe in the real world.

The first thing to check whether this motion that we have observed is linear. Some basic linear algebra texts don't bother to check this, but if you are going to try to find a matrix that represents a transformation, *you must*, one way or another, first check that the transformation is linear. In fact, for any function $f : \mathbb{R}^n \to \mathbb{R}^m$ you can go ahead and find $S = \{f(\mathbf{e}_1), \ldots, f(\mathbf{e}_n)\}$ and create the matrix $\mathbf{C}_S$ as you did in the previous section. But if the original function f is *not* linear, it would be false that $f = T_{\mathbf{C}_S}$. Put another way, if you try this process with a function that is not linear, almost all computations that you make with $T_{\mathbf{C}_S}$ would be *wrong*.

In MP 56 it was easy enough to check that the transformation was linear—you could just check the linearity of the components of T. But rotation has so far in this text been described *geometrically*, and therefore we must check *geometrically* that rotations are linear. We have to check two things about ρ_θ for scalars k and vectors $\mathbf{v}$ and $\mathbf{w}$: $\rho_\theta(k\mathbf{v}) = k\rho_\theta(\mathbf{v})$ and

$\rho_\theta(\mathbf{v} + \mathbf{w}) = \rho_\theta(\mathbf{v}) + \rho_\theta(\mathbf{w})$, using your geometric understanding of what a "rotation" is. It should be clear that if you multiply a vector $\mathbf{v}$ by a scalar k (which changes its length and possibly flips it) and rotate it to obtain $\rho_\theta(k\mathbf{v})$, then you will get the same vector as if you rotate $\mathbf{v}$ first and multiply it by the scalar k, which is the geometric way of saying that $\rho_\theta(k\mathbf{v}) = k\rho_\theta(\mathbf{v})$. Likewise, if you rotate $\mathbf{v} + \mathbf{w}$ (which points to the opposite corner of the parallelogram spanned by them) to get $\rho_\theta(\mathbf{v} + \mathbf{w})$, then because the parallelogram is rigidly rotated, you will get the same vector as if you first rotate $\mathbf{v}$ and $\mathbf{w}$, and then add them. That is, $\rho_\theta(\mathbf{v} + \mathbf{w}) = \rho_\theta(\mathbf{v}) + \rho_\theta(\mathbf{w})$

To reiterate here. We are trying to mathematically describe something we have observed, which we call rotation by angle θ, ρ_θ. We have checked that rotation, according to our geometric description, is linear. Having done so, we can now find the matrix that represents it. And to do so, we need only check what happens to $\mathbf{e}_1$ and $\mathbf{e}_2$, and use $\rho_\theta(\mathbf{e}_1)$ and $\rho_\theta(\mathbf{e}_1)$ as the columns of a 2×2 matrix. If we rotate $\mathbf{e}_1$ by angle θ, then elementary geometry shows us that $\mathbf{e}_1$ goes to the vector $(\cos\theta, -\sin\theta)$, and $\mathbf{e}_2$ goes to $(\sin\theta, \cos\theta)$. Therefore, the matrix representing ρ_θ is

$$\mathbf{R}_\theta = \begin{bmatrix} \cos\theta & \sin\theta \\ -\sin\theta & \cos\theta \end{bmatrix} \tag{4.29}$$

We can now make short work of the original problem of rotating $(1,3)$ by angle $\frac{\pi}{3}$. Using the geometry of a 30-60-90 triangle, $\cos\frac{\pi}{3} = \frac{1}{2}$ and $\sin\frac{\pi}{3} = \frac{\sqrt{3}}{2}$, which gives us the matrix

$$\begin{bmatrix} \frac{1}{2} & \frac{\sqrt{3}}{2} \\ -\frac{\sqrt{3}}{2} & \frac{1}{2} \end{bmatrix}$$

and multiplying this matrix by $\begin{bmatrix} 1 \\ 3 \end{bmatrix}$ gives us the point $\left(\frac{1+3\sqrt{3}}{2}, \frac{3-\sqrt{3}}{2}\right)$. And having done the work to find the matrix $\mathbf{R}_\theta$ *one time*, you can now just ask a computer to compute the matrix for a specific angle, then apply that matrix to *any* point in the plane.

Here are a few more observations to be gained from the matrix $\mathbf{R}_\theta$. Importantly, we are using different notation for the matrix $\mathbf{R}_\theta$ and the transformation ρ_θ that it represents. Geometrically speaking, you know that you should be able to "undo" a rotation of angle θ by rotating by angle $-\theta$. This suggests that $\mathbf{R}_\theta$ should be invertible, and that $\mathbf{R}_\theta^{-1} = \mathbf{R}_{-\theta}$. You will check this explicitly in the next exercise.

Exercise 4.14 *Show that $\mathbf{R}_\theta\mathbf{R}_{-\theta} = \mathbf{I}$. That is, $\mathbf{R}_\theta$ and $\mathbf{R}_{-\theta}$ are inverses. You will need some trigonometric identities. In case you have forgotten them, there is a nice list on Wikipedia, including things like $\cos(-\theta) = \cos(\theta)$.*

Exercise 4.15 *In this exercise you will do the reverse. You know that geometrically speaking, you should agree that rotating by angle θ_1 followed by rotation by angle θ_2 results in a rotation of angle $\theta_1 + \theta_2$. Compute the product matrix $\mathbf{R}_{\theta_2}\mathbf{R}_{\theta_1}$.*

1. *The upper left corner of the matrix should be equal to* $\cos(\theta_1 + \theta_2)$. *Look up the angle sum formula for cosine (online or in a calculus book) and confirm that your entry is in fact equal to* $\cos(\theta_1 + \theta_2)$.

2. *Do the same for the upper right entry.*

3. *Explain from your answer why rotation matrices commute (you don't need to, and shouldn't, multiply the matrices in the opposite order to show this).*

4.8 Composition of Linear Transformations

In the previous section you worked with the idea of rotating by angle θ_1 and rotating again by angle θ_2. This meant first multiplying on the left by $\mathbf{R}_{\theta_1}$, followed by $\mathbf{R}_{\theta_2}$ to the vector $\mathbf{v}$. Using the language of transformations, the result on a vector $\mathbf{v}$ is $T_{\mathbf{R}_{\theta_2}}(T_{\mathbf{R}_{\theta_1}}(\mathbf{v}))$. You have seen this kind of notation before—for example, when you learned about the chain rule to find derivatives of functions like $h(x) = \sin(x^2)$. This function is a *composition* of the functions $f(x) = \sin x$ and $g(x) = x^2$. So $h(x) = f(g(x))$, and you also write $h = f \circ g$. Early in calculus you likely only considered compositions of real functions—that is, the domain and codomain were always $\mathbb{R}$. Later in calculus you may have considered more general transformations and their compositions, although they may not have been described as compositions of functions. You should now recognize that the expression $T_{\mathbf{R}_{\theta_2}}(T_{\mathbf{R}_{\theta_1}}(\mathbf{v}))$ is really a composition of functions $T_{\mathbf{R}_{\theta_2}} \circ T_{\mathbf{R}_{\theta_1}}$ where each of the two functions has domain and codomain $\mathbb{R}^2$. However, it is not required that functions have the same domain and codomain in order to take their composition. For example, if you have a transformation $T_1 : \mathbb{R}^n \to \mathbb{R}^m$ and another transformation $T_2 : \mathbb{R}^m \to \mathbb{R}^k$, then T_1 takes an n-vector $\mathbf{v}$ and gives back an m-vector $T_1(\mathbf{v})$. Since $T_1(\mathbf{v})$ is an m-vector it is in the domain of T_2, and so we can apply T_2 to obtain $T_2(T_1(\mathbf{v}))$, which is a k-vector.

The point here is that as long as the codomain of T_1 is contained in the domain of T_2, then $T_2(T_1(\mathbf{v}))$ is defined, and is a vector in the range of T_2. In summary, if $T_1 : \mathbb{R}^n \to \mathbb{R}^m$ and $T_2 : \mathbb{R}^m \to \mathbb{R}^k$ then the composition $T_2 \circ T_1 : \mathbb{R}^n \to \mathbb{R}^k$ is defined by $(T_2 \circ T_1)(\mathbf{v}) = T_2(T_1(\mathbf{v}))$. If T_1 and T_2 are linear, then according to Theorem 4.11 there are matrices $\mathbf{A}_1$ and $\mathbf{A}_2$ such that $T_1 = T_{\mathbf{A}_1}$ and $T_2 = T_{\mathbf{A}_2}$. Let's now compute $T_2 \circ T_2$ in terms of these matrices:

$$T_2 \circ T_1(\mathbf{v}) = T_2(T_1(\mathbf{v})) = T_{\mathbf{A}_2}(T_{\mathbf{A}_1}(\mathbf{v})) = \mathbf{A}_2(\mathbf{A}_1\mathbf{v}) = (\mathbf{A}_2\mathbf{A}_1)\mathbf{v} = T_{\mathbf{A}_2\mathbf{A}_1}(\mathbf{v})$$

We have proved the following theorem.

Theorem 4.12 *If T_1 and T_2 are linear, then $T_2 \circ T_1$ is linear. Moreover, if $T_1 = T_{\mathbf{A}_1}$ and $T_2 = T_{\mathbf{A}_2}$ then $T_2 \circ T_1 = T_{\mathbf{A}_2 \mathbf{A}_1}$.*

More briefly, *composition* of linear transformations corresponds to *multiplication* of the corresponding matrices. The is the single most important reason why multiplication of matrices is defined in the way that it is.

MP 57 *MATRIX OF A COMPOSITION*

The last thing to consider in this section is the inverse of a linear transformation $T : \mathbb{R}^n \to \mathbb{R}^n$. The inverse T^{-1}, assuming it exists, is also a function from $\mathbb{R}^n \to \mathbb{R}^n$, which, informally speaking "undoes whatever T does". You have seen inverses already in calculus—for example, the inverse of the function $f(x) = e^x$ is the natural log function function $g(x) = \ln x$. This is because the composition of these functions takes you back where you started: $e^{\ln x} = x$ and $\ln(e^x) = x$. The corresponding statement for transformations is $T \circ T^{-1}(\mathbf{v}) = \mathbf{v}$ and $T^{-1} \circ T(\mathbf{v}) = \mathbf{v}$. If we suppose that $\mathbf{A}$ is the matrix corresponding to T and $\mathbf{B}$ is the matrix corresponding to T^{-1}, then $T \circ T^{-1}(\mathbf{v}) = \mathbf{v}$ translates into

$$\mathbf{A}\mathbf{B}\mathbf{v} = \mathbf{v}$$

As you know, the identity matrix has the property that $\mathbf{I}\mathbf{v} = \mathbf{v}$ for every vector $\mathbf{v}$. According to the uniqueness of Theorem 4.10, this means that $\mathbf{A}\mathbf{B} = \mathbf{I}$, which in turn means that $\mathbf{B} = \mathbf{A}^{-1}$. That is, if T has an inverse then its corresponding matrix $\mathbf{A}$ is invertible, and the transformation $T_{\mathbf{A}^{-1}}$ is the equal to T^{-1}. Likewise, if T is not invertible then its matrix $\mathbf{A}$ cannot have an inverse. We summarize this discussion in the final theorem of this section:

Theorem 4.13 *A linear transformation $T : \mathbb{R}^n \to \mathbb{R}^n$ has an inverse if and only if the matrix $\mathbf{A}$ corresponding to T has an inverse. If T does have an inverse then the matrix corresponding to T^{-1} is $\mathbf{A}^{-1}$.*

4.9   Geometric Meaning of the Determinant

Consider two non-zero vectors $\mathbf{u} = (u_1, 0)$ and $\mathbf{v} = (v_1, v_2)$ in the plane, where u_1, v_1, v_2 are all positive. Then $\mathbf{u}$ and $\mathbf{v}$, together with the translate of $\mathbf{u}$ to the end of $\mathbf{v}$ and the translate of $\mathbf{v}$ to the end of $\mathbf{u}$ form the sides of a parallelogram P in the first quadrant of the plane. If the $\mathbf{u}$ and $\mathbf{v}$ are parallel, then the parallelogram is *degenerate* in the sense that its sides are all collinear. In this case we sensibly say that the area of P is 0. Otherwise, let's calculate the area of P. You can look up a formula for this of course, but that's missing the point. Geometrically speaking, the parallelogram is composed of a "tilted stack" of segments with length $\|\mathbf{u}\| = u_1$, with one for each y value from 0 to v_2. If you have studied areas by slices in calculus, this idea should be familiar to you. At any rate, the area A of the resulting rectangle, and therefore the original parallelogram P, is equal to $u_1 v_2$. One way to see this is to imagine sliding all of those slices horizontally until they form a rectangle, which has the same height as P the same width of the segments. This formula is also equal to the formula you may have learned or looked up on the internet, that the area of a parallelogram is the base times the height, which can also be derived using elementary Euclidean geometry. On the other hand, $\det \begin{bmatrix} u_1 & v_1 \\ 0 & v_2 \end{bmatrix} = u_1 v_2$. For a degenerate parallelogram, the vectors in question are scalar multiples of one another, and you know already that the matrix containing them as rows has determinant 0. This shows that for the parallelogram P, spanned by vectors $S = \{\mathbf{u}, \mathbf{v}\}$ situated in this way in the plane, the area of P is $\det \mathbf{C}_S$.

Now suppose P is any parallelogram in the plane spanned by vectors $\mathbf{u}$ and $\mathbf{v}$. The first thing we will do is rotate $\mathbf{u}$ until it is in the position described above. On the one hand, geometrically speaking, rotation doesn't change the area of the parallelogram. As you already learned, rotation is accomplished by applying a rotation matrix as defined in Formula (4.29). To see where the vectors $\mathbf{u}$ and $\mathbf{v}$ both go, we first put them as columns of a matrix

$$\mathbf{B} = \begin{bmatrix} u_1 & v_1 \\ u_2 & v_2 \end{bmatrix}$$

multiply on the left by the rotation matrix $\mathbf{R}_\theta = \begin{bmatrix} \cos\theta & \sin\theta \\ -\sin\theta & \cos\theta \end{bmatrix}$. We choose θ so that the original vector $\mathbf{u}$ is rotated to a new vector $\mathbf{w} = (w_1, 0)$. Does this change the determinant of the matrix? As you know, the determinant of the product of matrices is the product of their determinants, so

$$\det(\mathbf{R}_\theta \mathbf{B}) = \det\left(\begin{bmatrix} \cos\theta & \sin\theta \\ -\sin\theta & \cos\theta \end{bmatrix} \mathbf{B}\right) = \left(\det \begin{bmatrix} \cos\theta & \sin\theta \\ -\sin\theta & \cos\theta \end{bmatrix}\right)(\det \mathbf{B})$$

$$= (\cos^2\theta + \sin^2\theta)(\det \mathbf{B}) = \det \mathbf{B}$$

In other words, *rotation does not change the determinant.* Now

$$\mathbf{R}_\theta \mathbf{B} = \begin{bmatrix} w_1 & z_1 \\ 0 & z_2 \end{bmatrix}$$

where $\mathbf{z} = (z_1, z_2) = \mathbf{R}_\theta(\mathbf{v})$.

After this rotation, the only remaining problem is that $\mathbf{z}$ might not be in the first quadrant–that is, one of its components (or both) might be negative. We can move $\mathbf{z}$ into the first quadrant by negating any negative components. In the next exercise you will draw these situations to convince yourself geometrically that negating the components doesn't impact the area of the parallelogram. But how does this change the determinant of the matrix? If $z_1 \geq 0$

and $z_2 < 0$, then we get the matrix

$$\begin{bmatrix} w_1 & z_1 \\ 0 & -z_2 \end{bmatrix}$$

from $\mathbf{R}_\theta \mathbf{B}$ by simply multiplying the bottom row by -1, which multiplies the determinant by -1. Likewise if we need to negate both z_1 and z_2 we can just multiply the second column by -1, which also changes the determinant by -1. The final case is an exercise.

Exercise 4.16 *Let* $\mathbf{u} = (0, 4)$. *In each case below, draw the vectors* $\mathbf{u}$, $\mathbf{v}$, *and* $\mathbf{v}'$, *where* $\mathbf{v}'$ *is obtained from* $\mathbf{v}$ *by negating all of its negative components. Sketch the parallelograms spanned by* $\mathbf{u}$ *and* $\mathbf{v}$ *and* $\mathbf{u}$ *and* $\mathbf{v}'$ *to see geometrically that they have the same area.*

    1. $\mathbf{v} = (2, -3)$ *(so* $\mathbf{v}' = (2, 3)$*).*

    2. $\mathbf{v} = (-2, -3)$

    3. $\mathbf{v} = (-2, 3)$

Exercise 4.17 *Consider the case when* $z_1 < 0$ *and* $z_2 \geq 0$ *in the above argument, meaning that in order to get* $\mathbf{z}$ *into the first quadrant we need to replace* z_1 *by* $-z_1$. *There are two cases to consider. If* $z_2 = 0$, *then* $\mathbf{z}$ *points in the opposite direction of* $\mathbf{w}$, *so the determinant is 0 and the parallelogram is degenerate, and has area 0. If* $z_2 > 0$ *explain how to change* z_1 *to* $-z_1$ *using a single row operation that does not change the determinant and **does not change** w_1.*

Putting all of this together, we see that any parallelogram spanned by two vectors can be rigidly moved into the position in which one of the vectors lies in the x-axis and both vectors are in the first quadrant, without changing its area. When doing so, the determinant of the matrix having those vectors as columns is changed by a factor of ± 1. This calculation, together with the fact that the transpose operation does not change the determinant (Theorem 3.8) shows:

Theorem 4.14 *If* $\mathbf{B}$ *is a* 2×2 *matrix then* $|\det \mathbf{B}|$ *is the area of the parallelogram spanned by its column vectors (or row vectors) in the plane.*

We can now check this theorem in $\mathbb{R}^3$, in which case the object spanned by three vectors is called a *parallelepiped*. We can verify this in two ways. The first uses cofactor expansion, and works in all dimensions. The second uses the cross product.

As in the plane, we can use rigid motions of $\mathbb{R}^3$ to put the three vectors so that $\mathbf{u}$ and $\mathbf{v}$ lie in the (x, y)-plane with $\mathbf{u}$ parallel to the x-axis and $\mathbf{v}$ in the first quadrant of that plane, and so that $\mathbf{w}$ is in the first octant $(w_1, w_2, w_3 \geq 0)$. We are not yet in a position to check that these motions change the determinant of a matrix only by a factor of ± 1 (see Theorem 6.19) and therefore just assume that for now. We now need to check the theorem for three vectors $\mathbf{u} = (u_1, u_2, 0)$, $\mathbf{v} = (v_1, v_2, 0)$, and $\mathbf{w} = (w_1, w_2, w_3)$ with all of the components non-negative. Then w_3 is the height of the parallelepiped above the x, y-plane. Again thinking of volume in terms of slices we see that the volume V of the parallelepiped is w_3 times the area of the parallelogram spanned by $\mathbf{u}$ and $\mathbf{v}$, which from our previous computation is

$$V = w_3 \left| \det \begin{bmatrix} u_1 & u_2 \\ v_1 & v_2 \end{bmatrix} \right|$$

On the other hand, let's drop all of these vectors as columns into a 3×3 matrix:

$$\begin{bmatrix} w_1 & u_1 & v_1 \\ w_2 & u_2 & v_2 \\ w_3 & 0 & 0 \end{bmatrix}$$

We can compute the determinant of this matrix by cofactor expansion in the bottom row, to get

$$w_3 \det \begin{bmatrix} u_1 & u_2 \\ v_1 & v_2 \end{bmatrix}$$

Again, this argument can be carried out in any dimension, showing the following theorem:

Theorem 4.15 *If* $\mathbf{A}$ *is an* $n \times n$ *matrix, then* $|\det \mathbf{A}|$ *is the volume of the parallelepiped spanned by its column (or row) vectors.*

A mathematical purist might say that the above theorem should really be the *definition of volume* of a parallelepiped in higher dimensions and that the discussion preceding it is not a proof but a geometric justification for the definition, but this distinction is of little practical value for you at this stage.

How does this work in $\mathbb{R}^3$ using the cross product (which is likely the definition you would have seen in calculus)? First, we saw earlier in Equation (4.27) that the area of the parallelogram spanned by $\mathbf{u}$ and $\mathbf{v}$ is $\|\mathbf{u} \times \mathbf{v}\|$. Adding another vector $\mathbf{w}$ into the mix, we can compute the volume V of the resulting parallelepiped by slices, each of which has area $\|\mathbf{u} \times \mathbf{v}\|$ multiplied by the height of the parallelepiped, which is $\|\mathbf{w}\| |\cos \theta|$, where θ is the angle between $\mathbf{w}$ and the plane spanned by $\mathbf{u}$ and $\mathbf{v}$. This angle can be more precisely defined later involving projections, but the meaning should be geometrically clear. At any rate, from the Triple Scalar Product Formula (4.26), we can calculate

$$V = \|\mathbf{w}\| |\cos \theta| \|\mathbf{u} \times \mathbf{v}\| = |\mathbf{w} \cdot (\mathbf{u} \times \mathbf{v})| = \left| \det \begin{bmatrix} w_1 & w_2 & w_3 \\ u_1 & u_2 & u_3 \\ v_1 & v_2 & v_3 \end{bmatrix} \right|$$

The matrix above is obtained from the matrix having these vectors as columns

$$\begin{bmatrix} w_1 & u_1 & v_1 \\ w_2 & u_2 & v_2 \\ w_3 & u_3 & v_3 \end{bmatrix}$$

by just swapping rows and taking the transpose, each of which only impacts the determinant by ± 1. Therefore

$$V = \left| \det \begin{bmatrix} w_1 & u_1 & v_1 \\ w_2 & u_2 & v_2 \\ w_3 & u_3 & v_3 \end{bmatrix} \right|$$

5

Vector Spaces

In this chapter we will extend what you have learned about Euclidean spaces $\mathbb{R}^n$ to more general objects called *vector spaces*. While the spaces $\mathbb{R}^n$ are certainly very important, there are many other important vector spaces that occur in science and engineering. In addition, treating $\mathbb{R}^n$ "abstractly" as a finite dimensional vector space also helps understand and utilize $\mathbb{R}^n$ itself. The main limitation of our previous approach to $\mathbb{R}^n$ is that it we are limited to the coordinate expressions of vectors using their "standard bases", while in real applications there are often other coordinate expressions better suited to the problem at hand. The notion of "basis change" that we will develop later frees us from these coordinate limitations.

5.1   Vector Spaces and Subspaces

A *vector space* V is a set of objects called *vectors,* together with two operations, addition and scalar multiplication, that satisfy the exact same basic algebraic properties of $\mathbb{R}^n$ that you have already learned. That is, for scalars k, c and vectors $\mathbf{u}, \mathbf{v}, \mathbf{w}$:

1. (Closure) $\mathbf{u} + \mathbf{v}$ and $k\mathbf{v}$ are also in V.

2. (Commutativity) $\mathbf{u} + \mathbf{v} = \mathbf{v} + \mathbf{u}$

3. (Associativity) $\mathbf{u} + (\mathbf{v} + \mathbf{w}) = (\mathbf{u} + \mathbf{v}) + \mathbf{w}$

4. (Zero Vector) There is a vector called the 0-vector $\mathbf{0}$ such that $\mathbf{0} + \mathbf{u} = \mathbf{u}$

5. (Multiplication by scalar 0) $0 \cdot \mathbf{u} = \mathbf{0}$

6. (Multiplication by scalar 1) $1 \cdot \mathbf{u} = \mathbf{u}$

7. (Distributive Laws For Scalars) $k(\mathbf{u} + \mathbf{v}) = k\mathbf{u} + k\mathbf{v}$ and $(c + k)\mathbf{u} = c\mathbf{u} + k\mathbf{u}$

8. (Associative Law for Scalars) $(kc)\mathbf{v} = k(c\mathbf{v})$

One property that is sometimes included in the definition of vector space is "existence of negative vectors". However, as we will see now, this can be

DOI: 10.1201/9781003674368-5

derived from the above properties. As usual, we denote $(-1)\mathbf{u}$ by $-\mathbf{u}$. Then we calculate the necessary property of $-\mathbf{u}$. In fact, using only the properties above:

$$\mathbf{u} + (-\mathbf{u}) = 1\mathbf{u} + (-1)\mathbf{u} = (1 + (-1))\mathbf{u} = 0\mathbf{u} = \mathbf{0}$$

You should read the above line carefully, noting when something is a scalar vs a vector, and justifying each "=" using the eight properties above. You should not attempt to memorize those eight properties. What is important is not whether you can quickly list them all out, but whether you can use them in practice. For example, how many of the properties that uniquely define the real numbers, including properties involving order and completeness, can you list out quickly? Even if you can't write them all out quickly you, are still able to do the basic arithmetic on which they depend!

The most obvious example of a vector space is $\mathbb{R}^n$. Again, it is important to distinguish between $\mathbb{R}^n$ considered as a space of *points* $P(x_1, \ldots, x_n)$ and as a space consisting of *vectors* $\mathbf{v} = (v_1, \ldots, v_n)$. We have already defined notions of vector addition and scalar multiplication, and we have already checked all of the properties above.

Example 5.1 *Just in case you were wondering:* $\mathbb{R}$ *itself is a vector space—with the usual operation of addition, and "scalar multiplication" just being ordinary multiplication. Alternatively, we can write real numbers as* $[a]$, *with scalar multiplication as usual for matrices.*

Example 5.2 *Let* $\mathcal{C}_{[0,2\pi]}$ *be the set of continuous functions* $f : [0, 2\pi] \to \mathbb{R}$. *You will already have seen in calculus the theorems necessary to show that* $\mathcal{C}$ *is a vector space, where the sum of* f *and* g *is defined by* $(f+g)(x) = f(x)+g(x)$ *and* $(kf)(x) = k(f(x))$. *Closure simply means that the sum and scalar multiples of continuous functions are continuous, which again are theorems from calculus. You may not have seen proper proofs of these theorems in calculus, which are usually found in an undergraduate course in Real Analysis. From the commutative property of addition of the real numbers, you know that* $f(x) + g(x) = g(x) + f(x)$, *showing the second property. This should convince you that properties 3 and 5-7 similarly follow from the algebraic properties of the real numbers, and I'll leave it up to you to check them. What should the 0-vector be? Clearly the function defined by* $f(x) = 0$ *for all* x *satisfies the required properties, as you can quickly check.*

Example 5.3 *We need not use* $[0, 2\pi]$ *as the domain in the above example—in fact any interval may be used, including* $(-\infty, \infty)$, *which is just all of* $\mathbb{R}$. *That is, the set* $\mathcal{C}$ *of all continuous functions* $f : \mathbb{R} \to \mathbb{R}$ *is also a vector space with the same operations of addition and scalar multiplication.*

Example 5.4 *Rather than n-tuples* $(x_1, \ldots, x_n)$ *of real numbers, which represent vectors in* $\mathbb{R}^n$, *one could just as well consider the set* $\mathbb{R}^\infty$ *of all infinite sequences* $(x_1, x_2, \ldots.)$ *of real numbers, and define scalar multiplication and addition as we did in* $\mathbb{R}^n$: $k(x_1, x_2, \ldots) = (kx_1, kx_2, \ldots)$ *and*

$(x_1, x_2, \ldots) + (y_1, y_2, \ldots) = (x_1 + y_1, x_2 + y_2, \ldots)$—*which is called "termwise addition". As you can imagine, essentially the same arguments that show that* $\mathbb{R}^n$ *is a vector space show that* $\mathbb{R}^\infty$ *is a vector space, and I'll leave the details to you if you choose to work them out.*

Example 5.5 *For this example we come full circle. We derived the algebraic properties of vectors by considering them as matrices. On the other hand, matrices can also be considered as vectors, forming the vector space* $\mathbb{M}_{m,n}$ *of all* $m \times n$ *matrices. You already know how to add matrices (which must be the same size) and multiply matrices by scalars, and the fact that* $\mathbb{M}_{m,n}$ *is a vector space follows from the algebraic properties of matrices that we proved early on in this course.*

By now you have noticed a pattern. The vector space properties in all of the examples above are essentially consequences of the basic properties of the real numbers. This is very typical—if something is a vector space, often it is not hard to actually prove that it is. Proving that something is *not* a vector space is often also straightforward because to show that something is *not* a vector space you have only to show that a *single one* of Properties 1-8 doesn't hold. And it is often quite easy to write down things that don't behave properly. In the following example and exercises we will use "⊞" to represent a possible addition operation, and I (or you) will check whether it actually satisfies all of the requirements to be a vector space.

Example 5.6 *On the set* $\mathbb{R}^2$, *define* $(x, y) \boxplus (x', y') = (x+x', y-y')$ *and scalar multiplication in the usual way. To show this is not a vector space, we need only find a single property among 1-8 that is not satisfied for a single choice of vectors or scalars. Going down the list and checking, you can stop right away at #2. According to our method of adding vectors,* $(1, 2) \boxplus (1, 3) = (2, -1)$, *but* $(1, 3) \boxplus (1, 2) = (2, 1)$, *so addition is not commutative.*

Exercise 5.1 *On* $\mathbb{R}^2$

> *1. Keep the definition of scalar multiplication but change the definition of vector addition to something new (other than what was done above) in such a way that at least one of Properties 1-8 is not satisfied (show which one). This is not as hard as it may sound—almost anything sightly weird will do it!*

> *2. Do the reverse of the above: Keep vector addition the same but change scalar multiplication to something so that one of Properties 1-8 is not satisfied.*

Exercise 5.2 *Define operations on* 2×2 *matrices by* $\mathbf{A} \boxplus \mathbf{B} = \mathbf{0}_{2\times 2}$ *with scalar multiplication defined as usual. State which of the properties of a vector space are true for these operations, where the "zero vector" is the matrix* $\mathbf{I}$. *No proof is necessary, but think carefully to be sure that you are correct!*

Exercise 5.3 *For 2×2 matrices, define $\mathbf{A} \boxplus \mathbf{B} = \mathbf{AB}$. That is, we are using matrix multiplication for our addition operation, with scalar multiplication as usual. State which of the properties of a vector space are true for these operations, where the "zero vector" is the matrix $\mathbf{I}$. No proof is necessary, but think carefully to be sure that you are correct!*

Many of the vector spaces that you will encounter are contained in other vector spaces.

Definition 5.1 *A non-empty subset S of a vector space V is called a subspace of V if with the operations $+$ and scalar multiplication of V, S is itself a vector space.*

Example 5.7 *Since (as you know from calculus) every polynomial is continuous, the set $\mathcal{P}$ of all polynomials $p : \mathbb{R} \to \mathbb{R}$ is a subset of the vector space $\mathcal{C}$ from Example 5.3. Recall that polynomials (in one variable) are functions of the form $p(x) = a_n x^n + a_{n-1} x^{n-1} + \ldots + a_1 x + a_0$. Written in summation notation,*

$$p(x) = \sum_{i=0}^{n} a_i x^i.$$

The numbers a_i are called the coefficients of p, and we always assume that $a_n \neq 0$ unless $n = 0$. The number n is called the degree of p. A constant function $p(x) = a_0$ is a degree 0 polynomial, even if $a_0 = 0$. Is $\mathcal{P}$ is a vector space in its own right with those operations? Let's start checking. As we will see later, the first property—Closure—is the crucial one. But if $q(x) = b_m x^m + b_{m-1} x^{m-1} + \cdots + b_1 x + b_0$ is another polynomial, then

$$p(x) + q(x) = \left(a_n x^n + a_{n-1} x^{n-1} + \cdots + a_1 x + a_0 \right)$$
$$+ \left(b_m x^m + b_{m-1} x^{m-1} + \cdots + b_1 x + b_0 \right)$$

Now it is possible that the degree m of q is not the same as the degree n of p, but we can still combine terms to get another polynomial of degree at most equal to the maximum of m and n. For a specific example, let $p(x) = 3x^3 + x - 4$ and $q(x) = 2x^2 - x$. Then

$$p(x) + q(x) = 3x^3 + 2x^2 - 4$$

Likewise, the scalar multiple of a polynomial is a polynomial. Looking at the rest of the properties of a vector space, it is easy to check that all of them are true, including the fact that the 0-function is a polynomial. I won't ask you to check them all because we will later see that Closure is the only one that matters. The vector space of polynomials is extremely important. In fact, polynomials have formulas that only involve addition and multiplication of numbers, and hence may be easily evaluated by computers. For this reason, approximation of functions by polynomials is an important method to understand functions in general.

Now suppose that we have a vector space V, and a non-empty subset S of V. We say that S is *closed under addition and scalar multiplication* if Closure is satisfied for S. Similar to what we already observed for polynomials, then Properties 2-3 and 5-8 are automatically true for vectors in S. Therefore the only remaining property to check to ensure that S is a subspace of V is whether the vector $\mathbf{0}$ is in S. But there is some vector $\mathbf{v}$ in S (because S is not empty) and by closure and Property 5, the scalar multiple $0\mathbf{v} = \mathbf{0}$ must be in $\mathbf{S}$. We state this observation as a theorem.

Theorem 5.1 *If S is a non-empty subset of a vector space V that is closed under addition and scalar multiplication, then S is a subspace of V.*

Remark 5.1 *By definition, closure means checking two conditions. (1) If $\mathbf{v}$ is in S and k is a scalar, then $k\mathbf{v}$ is in S. (2) If $\mathbf{v}$ and $\mathbf{w}$ are in S, then $\mathbf{v} + \mathbf{w}$ is in S. Often (but not always!) it is more efficient to prove these two properties simultaneously by showing that if k is a scalar and $\mathbf{v}, \mathbf{w}$ are in S, then $k\mathbf{v} + \mathbf{w}$ is in S. If this has been proved, then we can obtain (1) by just letting $\mathbf{w} = \mathbf{0}$, and (2) by just taking $k = 1$. We will often employ this strategy, and so should you!*

To repeat, the above theorem saves you a lot of work: rather then checking all the Properties 1-8 to see if a subset of a vector space is a vector space in its own right, you only must check the first condition.

Example 5.8 *The space $\mathcal{F}$ of all functions (continuous or not) $f : \mathbb{R} \to \mathbb{R}$ is a vector space. The argument is exactly the same as the one we used with continuous functions but simpler: the sum of functions is a function, and the scalar multiple of functions is a function, with no need to worry about continuity. The space $\mathcal{F}$ has many important subspaces, including $\mathcal{C}$. From calculus you will have learned that differentiable functions are continuous, and also that the sum and scalar multiple of differentiable functions are differentiable. That is, the closure property is satisfied for the set $\mathcal{D}$ of differentiable functions, meaning it is a subspace of $\mathcal{C}$. We already checked that $\mathcal{P}$ is a subspace of $\mathcal{C}$, but since polynomials are also differentiable, $\mathcal{P}$ is also a subspace of $\mathcal{D}$. Finally, we can let $\mathcal{P}(n)$ denote the set of all polynomials of degree $\leq n$. Then $\mathcal{P}(n)$ is a subspace of $\mathcal{P}$ because the sum of polynomials has degree at most equal to the maximum of the degrees of the two polynomials, and scalar multiplication doesn't change the degree of a polynomial unless the scalar is 0—in which case it changes the degree to 0.*

Summarizing the above example, $\mathcal{P}(n)$ is a subspace of $\mathcal{P}$, which is a subspace of $\mathcal{D}$, which is a subspace of $\mathcal{C}$, which is a subspace of $\mathcal{F}$.

Exercise 5.4 *Show that the set of polynomials of degree **equal** to n is not a subspace of $\mathcal{P}$ for $n = 2$ by finding two polynomials of degree 2 whose sum does not have degree 2. Of course there are similar examples for every n.*

Example 5.9 *Subspaces often arise by restricting to smaller sets that have a specific property that is closed under addition and scalar multiplication. For example, the sum of two lower triangular matrices is lower triangular, as is the scalar product of a lower triangular matrix. This means that the set of lower triangular matrices is a subspace of the vector space $\mathbb{M}_{n,n}$ of $n \times n$ matrices.*

Exercise 5.5 *Show that the subset of $n \times n$ matrices with determinant 0 is **not** a subspace, of $\mathbb{M}_{n,n}$, for $n = 2$. That is, find two 2×2 matrices of determinant 0 whose sum does not have determinant 0. Of course similar examples exist for any n.*

Example 5.10 *The set of all 2×3 matrices of the form $\begin{bmatrix} 0 & x & 2x \\ 0 & 0 & 0 \end{bmatrix}$, where x is a real number, is a subspace of $\mathbb{M}_{2,3}$. We will check closure using Remark 5.1. Let $\mathbf{A} = \begin{bmatrix} 0 & a & 2a \\ 0 & 0 & 0 \end{bmatrix}$ and $\mathbf{B} = \begin{bmatrix} 0 & b & 2b \\ 0 & 0 & 0 \end{bmatrix}$ be two matrices of this form, and k be a scalar. Then*

$$
k\mathbf{A} + \mathbf{B} = k \left(\begin{bmatrix} 0 & a & 2a \\ 0 & 0 & 0 \end{bmatrix} + \begin{bmatrix} 0 & b & 2b \\ 0 & 0 & 0 \end{bmatrix} \right)
$$

$$
= \begin{bmatrix} 0 & k(a+b) & 2\left[k(a+b)\right] \\ 0 & 0 & 0 \end{bmatrix}
$$

which is of the required form.

On the other hand, matrices of the form $\begin{bmatrix} 1 & x & 2x \\ 0 & 0 & 0 \end{bmatrix}$ do not form a subspace of $\mathbb{M}_{2,3}$. For example,

$$
2 \begin{bmatrix} 1 & x & 2x \\ 0 & 0 & 0 \end{bmatrix} = \begin{bmatrix} 2 & 2x & 4x \\ 0 & 0 & 0 \end{bmatrix}
$$

is not of the required form, so closure of scalar multiplication fails.

Exercise 5.6 *Show that closure of the sum also fails for matrices of the form $\begin{bmatrix} 1 & x & 2x \\ 0 & 0 & 0 \end{bmatrix}$. Note that to verify that a set S is **not** a subspace, you need only check that either closure of scalar multiplication fails, or closure of the sum fails. In some cases, one may fail but not the other, but this still means that S is not a subspace.*

Exercise 5.7 *Show whether 2×2 matrices of the following forms are a subspace of the vector space of $\mathbb{M}_{2,2}$, where x and y are real numbers. Hint: To show they are a subspace, you need to show that sums and scalar multiples can be put into the same general form. To show they are not, you just need to show that Closure fails for specific matrices (or if it is simpler, you can show that any of Properties 1-8 fails).*

1. $\begin{bmatrix} x & 0 \\ y & x+y \end{bmatrix}$

2. $\begin{bmatrix} 1 & 0 \\ y & x+y \end{bmatrix}$

3. $\begin{bmatrix} x^2 & 0 \\ x & x \end{bmatrix}$

The next theorem provides a very important source of subspaces. You have no doubt seen Venn diagrams involving the intersection of two or three sets. The intersection of sets consists of the elements that are in *all* of the sets simultaneously. It is possible to take intersections of infinitely many sets, and the next theorem is also true for infinite intersections, but we will not need this fact and we will only prove it for finite intersections.

Theorem 5.2 *The intersection of subspaces of a vector space V is a subspace of V.*

Proof. Suppose that $S_1, \ldots, S_n$ are subspaces of V. Recall that the intersection of the sets $S_1, \ldots, S_n$ is the set of all $\mathbf{u}$ that are in *all* of the sets S_i simultaneously. Since all of the subspaces S_i contain $\mathbf{0}$ (Property 4), their intersection S contains $\mathbf{0}$, and S is not empty. We need to check that S is closed under addition and scalar multiplication. Suppose $\mathbf{u}$ and $\mathbf{v}$ are in S. Then $\mathbf{u}$ and $\mathbf{v}$ are in every S_i. Since every S_i is closed under addition, this means that $\mathbf{u} + \mathbf{v}$ is in every S_i. Therefore $\mathbf{u} + \mathbf{v}$ is in S. The last part of the proof is an exercise. ∎

Exercise 5.8 *Finish the proof of the above theorem by showing that if k is any scalar and $\mathbf{u}$ is in S then $k\mathbf{u}$ is in S.*

Exercise 5.9 *Show that the union of subspaces of a vector space V generally is not a subspace of V by exhibiting a counterexample to the statement. Recall that if S_1 and S_2 are sets, the union $S_1 \cup S_2$ consists of all s that are in S_1 or in S_2. Do so as follows: As in Example 5.9, the set L of lower triangular matrices is a subspace of $\mathbb{M}_{2 \times 2}$. Likewise the set U of upper triangular matrices is a subspace of $\mathbb{M}_{2 \times 2}$. The union of U and L consists of all matrices that are either upper triangular or lower triangular (or both). Now write down an upper triangular matrix and a lower triangular matrix whose sum is neither. This verifies that closure of the sum fails for $U \cup L$ (but as you can check, closure of scalar multiplication is true).*

5.2 Linear Transformations

The exact same definition of linear transformation for Euclidean spaces works as well for any vector space.

Definition 5.2 *If V and W are vector spaces, a function $T : V \to W$ is called a linear transformation if for all $\mathbf{u}, \mathbf{v}$ in V and scalar k, $T(k\mathbf{u} + \mathbf{v}) = kT(\mathbf{u}) + T(\mathbf{v})$. The kernel of T is the set of all $\mathbf{v}$ in T such that $T(\mathbf{v}) = \mathbf{0}$. The range or image of T is the set of all $\mathbf{w}$ in W such that there is some $\mathbf{u}$ in V such that $T(\mathbf{u}) = \mathbf{w}$.*

Exercise 5.10 *Show that if $T : V \to W$ is a linear transformation between vector spaces, then $T(\mathbf{0}) = \mathbf{0}$. That is $\mathbf{0}$ is always in the kernel of T. Hint: you've seen this kind of argument before.*

As a reminder, V is called the *domain* of T and W is called the *codomain*. Sometimes it is convenient to separate the conditions for a linear transformation into two equivalent conditions, as we have done in the past. That is, $T(k\mathbf{u}) = kT(\mathbf{u})$ and $T(\mathbf{u} + \mathbf{v}) = T(\mathbf{u}) + T(\mathbf{v})$. I will occasionally also split the proof up in this way–proving one condition and asking you to prove the other. The following is a very important theorem about linear transformations.

Theorem 5.3 *If $T : V \to W$ is a linear transformation between vector spaces, then $\ker T$ is a subspace of V and the range of T is a subspace of W.*

Proof. Suppose $\mathbf{v}$ and $\mathbf{w}$ are in $\ker T$ and k is a scalar. This means that $T(\mathbf{v}) = T(\mathbf{w}) = \mathbf{0}$. Using the linearity of T,

$$T(\mathbf{v} + \mathbf{w}) = T(\mathbf{v}) + T(\mathbf{w}) = \mathbf{0} + \mathbf{0} = \mathbf{0}$$

and

$$T(k\mathbf{v}) = kT(\mathbf{v}) = k\mathbf{0} = \mathbf{0}$$

This proves that the kernel is a subspace.

The range of T consists of all vectors $T(\mathbf{v})$ for $\mathbf{v}$ in V. Suppose that $T(\mathbf{v})$ and $T(\mathbf{w})$ are in the range of T. We need to check that $T(\mathbf{v}) + T(\mathbf{w})$ is in the range of T, meaning that $T(\mathbf{v}) + T(\mathbf{w}) = T(\mathbf{u})$ for some vector $\mathbf{u}$. The obvious choice is to just let $\mathbf{u} = \mathbf{v} + \mathbf{w}$. By the linearity of T,

$$T(\mathbf{u}) = \mathbf{T}(\mathbf{v} + \mathbf{w}) = \mathbf{T}(\mathbf{v}) + \mathbf{T}(\mathbf{w})$$

The rest of the proof is an exercise. ∎

Exercise 5.11 *Finish the proof of Theorem 5.3 by showing that for any scalar k and $T(\mathbf{v})$ in the range of T, $kT(\mathbf{v})$ is in the range of T.*

Theorem 5.4 *The set S of solutions to a homogeneous system $\mathbf{AX} = \mathbf{0}$ of linear equations is a subspace of $\mathbb{R}^n$.*

Proof. As we have already observed in Section 4.6, S is precisely the kernel of the linear transformation $T_{\mathbf{A}}$. ∎

Theorem 5.5 *If $T_1 : V \to W$ and $T_2 : W \to Z$ are linear transformations between vector spaces, then $T_2 \circ T_1 : V \to Z$ is a linear transformation.*

Proof. If **u** is in V and k is a scalar, then

$$(T_2 \circ T_1)(k\mathbf{u}) = T_2(T_1(k\mathbf{u})) = T_2(kT_1(\mathbf{u})) = kT_2(T_1(\mathbf{u})) = k(T_2 \circ T_1)(\mathbf{u})$$

The rest of the proof is an exercise. ∎

Exercise 5.12 *Finish the proof of the above theorem by showing that*

$$(T_2 \circ T_1)(\mathbf{u} + \mathbf{v}) = T_2 \circ T_1(\mathbf{u}) + T_2 \circ T_1(\mathbf{v})$$

5.3 Linear Combinations and Linear Independence

One of the most important ways to combine vectors is via *linear combinations*. That is, expressions of the form

$$k_1 \mathbf{v}_1 + \cdots + k_r \mathbf{v}_r$$

where $\mathbf{v}_1, \ldots, \mathbf{v}_r$ are vectors in a vector space V, and $k_1, \ldots, k_r$ are scalars. It is convenient to use summation notation for linear combinations. That is, the above sum can be more compactly written as

$$\sum_{i=1}^{r} k_i \mathbf{v}_i$$

You have already seen a linear combination in (4.28).

Given a set $S = \{\mathbf{v}_1, \ldots, \mathbf{v}_r\}$ of vectors in a vector space V, the *span* of S, denoted by *span* S or *span* $\{\mathbf{v}_1, \ldots, \mathbf{v}_n\}$ is the set of all linear combinations $\sum_{i=1}^{r} k_i \mathbf{v}_i$. If $S = \{\mathbf{v}\}$ is a single vector, then a linear combination of S is of the form $k_1 \mathbf{v} + \cdots + k_r \mathbf{v}$. But we can "collect terms" (formally, using the distributive law for scalars) to write $k_1 \mathbf{v} + \cdots + k_r \mathbf{v} = (k_1 + \cdots k_r)\mathbf{v}$, which is simply a scalar multiple of $\mathbf{v}$. That is, *span* $\{\mathbf{v}\}$ is the set of all scalar multiples $k\mathbf{v}$. As you have seen before, in $\mathbb{R}^n$, if $\mathbf{v} \neq \mathbf{0}$ then geometrically speaking *span* $\{\mathbf{v}\}$ consists of the line parallel to $\mathbf{v}$ through the origin.

In general, you can always use the distributive law to "collect terms" and therefore when writing a linear combination of a set of vectors $S = \{\mathbf{v}_1, \ldots, \mathbf{v}_r\}$ we always use each $\mathbf{v}_i$ *exactly once* in the sum. This includes the situation in which $\mathbf{v}_i = \mathbf{v}_j$ for some $i \neq j$–we will still consider $\mathbf{v}_i$ and $\mathbf{v}_j$ as distinct vectors in the sum.

There is one other simple observation that we will use below: each $\mathbf{v}_i$ is itself in *span* S because we may always write $\mathbf{v}_i$ as the linear combination $0\mathbf{v}_1 + \cdots + 1\mathbf{v}_i + \cdots + 0\mathbf{v}_r$. Another way to write this linear combination is

$$\sum_{j \neq i} 0\mathbf{v}_j + 1\mathbf{v}_i$$

Theorem 5.6 *If $S = \{\mathbf{v}_1, \ldots, \mathbf{v}_r\}$ is a set of vectors in a vector space V then span S is a subspace of V.*

Proof. Suppose $\mathbf{u} = \sum_{i=1}^{r} k_i \mathbf{v}_i$ and $\mathbf{w} = \sum_{i=1}^{r} c_i \mathbf{v}_i$ lie in *span* S, and k is a scalar. In order to show that $k\mathbf{u} + \mathbf{w}$ is in *span* S we need to write it as a linear combination of the vectors in S. Using the distributive and commutative laws for the real numbers we can calculate

$$k\mathbf{u} + \mathbf{w} = k\left(\sum_{i=1}^{r} k_i \mathbf{v}_i\right) + \sum_{i=1}^{r} c_i \mathbf{v}_i = \sum_{i=1}^{r} (kk_i + c_i)\,\mathbf{v}_i$$

■

It is an important possibility that a linear combination $k_1 \mathbf{v}_1 + \cdots k_n \mathbf{v}_n$ can be written as a linear combination of some proper subset of vectors in S. A fairly obvious possibility is that $\mathbf{v}_n = \mathbf{0}$. In that case

$$k_1 \mathbf{v}_1 + \cdots + k_n \mathbf{v}_n = k_1 \mathbf{v}_1 + \cdots + k_{n-1} \mathbf{v}_{n-1} + k_n \mathbf{0}$$

$$= k_1 \mathbf{v}_1 + \cdots + k_{n-1} \mathbf{v}_{n-1}$$

As soon as we drop the $\mathbf{0}$ term, the expression is no longer a linear combination of vectors in the original S, but rather of vectors in the smaller set $S' = \{\mathbf{v}_1, \ldots, \mathbf{v}_{n-1}\}$. Put another way, *span* S is equal to *span* S'. That is, if $\mathbf{0} = \mathbf{v}_i$ for some i, then simply dropping $\mathbf{v}_i$ from S does not change the span. Therefore, while we still theoretically allow $\mathbf{0}$ to theoretically be a vector in S, from a practical standpoint we can always remove $\mathbf{0}$ without changing the span of S, unless $S = \{\mathbf{0}\}$. As you will see later, it is vital to allow the case $S = \{\mathbf{0}\}$, but this is the only situation in which $\mathbf{0}$ plays an important role in linear combinations.

Now suppose that $S = \{\mathbf{v}_1, \mathbf{v}_2\}$ and there is some scalar k such that $\mathbf{v}_2 = k\mathbf{v}_1$. Then any linear combination $k_1 \mathbf{v}_1 + k_2 \mathbf{v}_2$ may we written as $(k_1 + kk_2)\,\mathbf{v}_1$. That is, *span* $\{\mathbf{v}_1, \mathbf{v}_2\}$ is equal to *span* $\{\mathbf{v}_1\}$. In some sense, $\mathbf{v}_2$ is "redundant" because $\mathbf{v}_2$ can be removed from S without changing the span. The formal term for this "redundancy" is *linear dependence*, which we will define below.

Now suppose $S = \{\mathbf{v}_1, \mathbf{v}_2, \mathbf{v}_3\}$. If, say, $\mathbf{v}_2 = k\mathbf{v}_1$, then similarly $\mathbf{v}_2$ is "redundant" because as in the above case, any linear combination of $\mathbf{v}_1, \mathbf{v}_2, \mathbf{v}_3$ can be written as a linear combination of $\mathbf{v}_1$ and $\mathbf{v}_3$. That is, *span* S is equal to *span* $\{\mathbf{v}_1, \mathbf{v}_3\}$. But the situation for three vectors is more subtle than the situation for two–"redundancy" can happen even when no vector is a scalar multiple of the other.

Exercise 5.13 *Consider the set $S = \{\mathbf{v}_1 = (0, 1, 1), \mathbf{v}_2 = (1, -1, 1), \mathbf{v}_3 = (2, -1, 3)\}$ in $\mathbb{R}^3$.*

> *1. Show that no two of these vectors is a scalar multiple of another. To get you started, I will show that $\mathbf{v}_3$ is not a scalar multiple of $\mathbf{v}_2$. If $\mathbf{v}_3 = k\mathbf{v}_2$ then since the first component of $\mathbf{v}_3$ is 2 and the first*

component of $\mathbf{v}_2$ is 1, it would have to be true that $k = 2$. Looking at the second components, this would mean that the second component of $\mathbf{v}_3$ would have to be $2(-1) = -2$, but instead it is -1. So there is no such number k.

2. Now check that $\mathbf{v}_1 + 2\mathbf{v}_2 = \mathbf{v}_3$.

3. Finally, show that you can replace any linear combination $k_1\mathbf{v}_1 + k_2\mathbf{v}_2 + k_3\mathbf{v}_3$ by a linear combination involving only $\mathbf{v}_1$ and $\mathbf{v}_2$. That is, $\mathbf{v}_3$ "redundant" even though no one of the three vectors was a scalar multiple of the others.

The above exercise motivates the following definition.

Definition 5.3 *A set $S = \{\mathbf{v}_1, \ldots, \mathbf{v}_r\}$ of vectors in a vector space is called linearly dependent if some $\mathbf{v}_i$ can be expressed as a linear combination of the others. If S is not linearly dependent, then S is called linearly independent.*

The same argument as in Exercise 5.13 works for any set $S = \{\mathbf{v}_1, \ldots, \mathbf{v}_r\}$ of vectors and we will basically repeat it in the proof of the next theorem, which formalizes the idea that linearly dependent sets contain "redundant" vectors. Recall that a set A is a *proper subset* of a set B if A is a subset of B but $A \neq B$. For example, $\{1, 2\}$ is a proper subset of $\{1, 2, 3\}$. Put another way, A is *strictly smaller* than B.

Theorem 5.7 *Let $S = \{\mathbf{v}_1, \ldots, \mathbf{v}_r\}$ be a set of vectors in a vector space. Then S is linearly dependent if and only if span S is equal to span S', where S' is a proper subset of S.*

Proof. Suppose S is linearly dependent. This means some $\mathbf{v}_j$ can be written as a linear combination of the other vectors in S. That is,

$$\mathbf{v}_j = c_1\mathbf{v}_1 + \cdots + c_{j-1}\mathbf{v}_{j-1} + c_{j+1}\mathbf{v}_{j+1} + \cdots + c_r\mathbf{v}_r \qquad (5.1)$$

or in summation notation:

$$\mathbf{v}_j = \sum_{i \neq j} c_i\mathbf{v}_i$$

Now suppose that $\sum_{i=1}^{r} k_i\mathbf{v}_i$ is in *span* S. Using the above summation we can write this as

$$\sum_{i=1}^{r} k_i\mathbf{v}_i = \sum_{i \neq j} k_i\mathbf{v}_i + k_j\mathbf{v}_j = \sum_{i \neq j} k_i\mathbf{v}_i + k_j \sum_{i \neq j} c_i\mathbf{v}_i = \sum_{i \neq j} (k_i + k_jc_i)\mathbf{v}_i$$

It is essential that you read the above equation carefully so that you understand how to work with the summation notation. In fact, it is basically the same calculation that you did in Exercise 5.13. To summarize, we started with a linear combination in *span* S and showed that it was actually a linear combination in S', where S' is obtained from S by *removing the linearly dependent vector* $\mathbf{v}_j$.

Conversely, suppose that *span S* is the same as *span S'*, when *S'* is obtained from *S* by removing a vector $\mathbf{v}_m$. As observed above, $\mathbf{v}_m$ is in *span S* and therefore is in *span S'*. By definition this means that

$$\mathbf{v}_m = \sum_{i \neq m} s_i \mathbf{v}_i$$

In other words, $\mathbf{v}_m$ is a linear combination of the other vectors in *S*. ∎

At this point we have two fundamental questions for a set $S = \{\mathbf{v}_1, \ldots, \mathbf{v}_r\}$:

Question 1: Is S linearly independent?

Question 2: Is a given vector $\mathbf{w}$ in span S?

In the rest of this section we will consider the first question.

Theorem 5.8 *A nonempty set $S = \{\mathbf{v}_1, \ldots, \mathbf{v}_r\}$ in a vector space V is linearly independent if and only if the only constants $k_1, \ldots, k_r$ that satisfy the equation $\sum_{i=1}^r k_i \mathbf{v}_i = \mathbf{0}$ are $k_1 = k_2 = \cdots = k_r = 0$. More conveniently, we say $k_i = 0$ for all i.*

Proof. Suppose there is a set of constants $k'_1, \ldots, k'_r$ with some $k'_j \neq 0$ such that $\sum_{i=1}^r k'_i \mathbf{v}_i = \mathbf{0}$. Pulling out the jth term, we can rewrite this equation as

$$\mathbf{0} = \sum_{i=1}^r k'_i \mathbf{v}_i = \sum_{i \neq j} k'_i \mathbf{v}_i + k'_j \mathbf{v}_j$$

Since $k'_j \neq 0$ we can multiply through by the scalar $\frac{1}{k'_j}$ and solve for $\mathbf{v}_j$

$$\mathbf{v}_j = \sum_{i \neq j} \frac{k'_i}{k'_j} \mathbf{v}_i$$

But this means that we have written $\mathbf{v}_j$ as a linear combination of the other vectors in *S*, so by definition *S* is linearly dependent.

Conversely, if we start by assuming *S* is linearly dependent, then we can essentially reverse this argument to obtain a linear combination $\sum_{i=1}^r c_i \mathbf{v}_i = \mathbf{0}$ with not all $c_i = 0$. The details are left as an exercise. ∎

Exercise 5.14 *Finish the proof of Theorem 5.8.*

In $\mathbb{R}^n$ we can relate linear independence to homogeneous systems of equations.

Theorem 5.9 *A nonempty set $S = \{\mathbf{v}_1, \ldots, \mathbf{v}_r\}$ in $\mathbb{R}^n$ is linearly independent if and only if the homogeneous system $\mathbf{C}_S \mathbf{X} = \mathbf{0}$ has a unique solution.*

Proof. By Theorem 5.8, *S* is linearly independent if and only if the only constants $k_1, \ldots, k_r$ that satisfy the equation $\sum_{i=1}^r k_i \mathbf{v}_i = \mathbf{0}$ are $k_i = 0$ for all i. Writing $\mathbf{v}_i = (v_{1i}, \ldots, v_{ni})$ we see that the equation $\mathbf{C}_S\mathbf{X} = \mathbf{0}$ may be written as

$$\begin{bmatrix} v_{11} & v_{12} & \cdots & v_{1r} \\ v_{21} & v_{22} & \cdots & v_{2r} \\ \vdots & \vdots & \vdots & \vdots \\ v_{n1} & v_{n2} & \cdots & v_{nr} \end{bmatrix} \begin{bmatrix} x_1 \\ x_2 \\ \vdots \\ x_r \end{bmatrix} = \begin{bmatrix} 0 \\ 0 \\ \vdots \\ 0 \end{bmatrix}$$

Now a vector $\begin{bmatrix} k_1 \\ k_2 \\ \vdots \\ k_r \end{bmatrix}$ is a solution to this equation if and only if $\sum_{i=1}^{r} k_i\mathbf{v}_i = \mathbf{0}$.

Moreover, the equation has a unique solution if and only if $\mathbf{X} = \mathbf{0}$ is the only solution, i.e. $k_i = 0$ for all i. ∎

The above theorem is the basis for the next MP:

MP 58 *LINEAR INDEPENDENCE*

5.4   Spanning Sets

Let's examine *Question 2*, whether a given vector $\mathbf{w}$ is in the span of a set $S = \{\mathbf{v}_1, \ldots, \mathbf{v}_r\}$. Put another way, can we find constants k_i and so that

$$\mathbf{w} = \sum_{i=1}^{r} k_i\mathbf{v}_i? \tag{5.2}$$

We will need to use double subscripts in what follows. That is, we will write $\mathbf{v}_i = (v_{i1}, \cdots, v_{in})$. In $\mathbb{R}^n$, writing the vectors as column vectors, the above equation is equivalent to

$$\begin{bmatrix} w_1 \\ w_2 \\ \vdots \\ w_n \end{bmatrix} = k_1 \begin{bmatrix} v_{11} \\ v_{12} \\ \vdots \\ v_{1n} \end{bmatrix} + k_2 \begin{bmatrix} v_{21} \\ v_{22} \\ \vdots \\ v_{3n} \end{bmatrix} + \cdots + k_r \begin{bmatrix} v_{r1} \\ v_{r2} \\ \vdots \\ v_{rn} \end{bmatrix}$$

Comparing entries on the left and right gives us $w_i = k_1v_{i1} + k_2v_{i2} + \cdots + k_rv_{rn}$. In other words, we can rewrite this equation in the form of a system of linear equations:

$$\begin{bmatrix} v_{11} & v_{21} & \cdots & v_{r1} \\ v_{21} & v_{22} & \cdots & v_{r2} \\ \vdots & \vdots & \vdots & \vdots \\ v_{1n} & v_{2n} & \cdots & v_{rn} \end{bmatrix} \begin{bmatrix} k_1 \\ k_2 \\ \vdots \\ k_r \end{bmatrix} = \begin{bmatrix} w_1 \\ w_2 \\ \vdots \\ w_n \end{bmatrix}$$

You should recognize that the matrix on the left is $\mathbf{C}_S$, and we may now rewrite this as $\mathbf{C}_S\mathbf{X} = \mathbf{W}$.

As you recall, inhomogeneous solutions may have a unique solution, infinitely many solutions, or no solution. In terms of answering the question at hand, we only care about whether this system has *any* solutions, i.e. whether the system is consistent. You have already learned how to answer this question, and you will practice in the next MP. The strategy is the same and only the phrasing of the equations has changed.

MP 59 *SPAN OF VECTORS*

We restate the above discussion as a theorem.

Theorem 5.10 *A vector* $\mathbf{w}$ *in* $\mathbb{R}^n$ *is a linear combination* $\mathbf{w} = \sum_{i=1}^{r} k_i\mathbf{v}_i$ *of vectors in the set* $S = \{\mathbf{v}_1, \ldots, \mathbf{v}_r\}$ *if and only if* $\begin{bmatrix} k_1 \\ k_2 \\ \vdots \\ k_r \end{bmatrix}$ *is a solution of the linear system*

$$\mathbf{C}_S\mathbf{X} = \mathbf{W} \tag{5.3}$$

In particular, $\mathbf{w}$ *is in the span of* S *if and only if the system (5.3) is consistent.*

MP 60 *EXPRESS AS LINEAR COMBINATION*

5.5   Bases

If V is a vector space, a set of vectors $B = \{\mathbf{v}_1, \ldots, \mathbf{v}_n\}$ is called a *basis* for V if B spans V and B is linearly independent. Roughly speaking, the fact that B spans V means that there are "enough" vectors in the set B that every $\mathbf{v}$ in V can be written as a linear combination $\mathbf{v} = \sum_{i=1}^{n} k_i\mathbf{v}_i$. On the other hand, linear independence of B means that there are "just enough" to do so. That is, according to Theorem 5.7, if we remove a single vector from B, then the remaining vectors no longer *span* V. The following theorem gives an equivalent way to show that a finite set is a basis for a vector space.

Theorem 5.11 *Let* $S = \{\mathbf{v}_1, \ldots, \mathbf{v}_n\}$ *be a subset of* V. *Then* S *is a basis for* V *if and only if every vector* $\mathbf{v}$ *in* V *can be written uniquely as a linear combination of vectors in* S. *Uniqueness means that if* $\mathbf{v} = \sum_{i=1}^{n} k_i\mathbf{v}_i = \sum_{i=1}^{n} m_i\mathbf{v}_i$ *then* $k_i = m_i$ *for all* i.

Proof. Suppose that S is a basis for V. Since S spans V, by definition every $\mathbf{v}$ in V can be written as a linear combination $\mathbf{v} = \sum_{i=1}^{n} k_i \mathbf{v}_i$. Suppose we also write $\mathbf{v} = \sum_{i=1}^{n} m_i \mathbf{v}_i$. Then

$$\mathbf{0} = \mathbf{v} - \mathbf{v} = \sum_{i=1}^{n} k_i \mathbf{v}_i - \sum_{i=1}^{n} m_i \mathbf{v}_i = \sum_{i=1}^{n} (k_i - m_i) \mathbf{v}_i$$

By Theorem 5.8, it must be true that $k_i - m_i = 0$ for all i, which means $k_i = m_i$ for all i, proving uniqueness.

In the converse direction, if every $\mathbf{v}$ may be written as a linear combination of vectors in S, then by definition S spans V. Since Theorem 5.8 is an if and only if statement, we may use the above argument in reverse, proving the theorem. ■

Example 5.11 *We have already seen the "standard basis" in $\mathbb{R}^n$, namely the vectors $\{\mathbf{e}_1, \ldots, \mathbf{e}_n\}$, where the components of $\mathbf{e}_i$ are all 0 except for the ith component, which is 1. Are these vectors linearly independent? Theorem 5.9 makes quick work of this question. If we make the matrix that has the vectors $\mathbf{e}_i$ as columns we get the identity matrix $\mathbf{I}$. The corresponding system has a unique solution because the $\det \mathbf{I} = 1 \neq 0$ (Theorem 3.10). The fact that the standard basis spans $\mathbb{R}^n$ was shown in (4.28).*

Example 5.12 *We define the "standard basis" for $\mathcal{P}(n)$ to consist of the polynomials $p_0(x) = 1$, $p_1(x) = x$, ..., $p_n(x) = x^n$. Then any polynomial of degree at most n may be written as a linear combination of these basic polynomials: $p(x) = \sum_{i=0}^{n} a_i x^i = \sum_{i=0}^{n} a_i p_i(x)$. By definition, this means that $\mathcal{P}(n)$ is spanned by the polynomials $p_0, \ldots, p_n$. These polynomials are also linearly independent. To see this, suppose that we have a linear combination $p(x) = \sum_{i=0}^{n} a_i p_i = \sum_{i=0}^{n} a_i x^i$ that is equal to the 0-polynomial. To show linear independence we need to show that $a_i = 0$ for all i. First, since $p(0) = a_0$ and $p(0) = 0$, $a_0 = 0$. Perhaps the simplest way to check that the rest of the coefficients are 0 is via calculus. Since $p(x) = 0$, $p'(x) = 0$, and*

$$p'(x) = n a_n x^{n-1} + (n-1) a_{n-1} x^{n-2} + \cdots + 2 a_2 x + a_1$$

Since $p'(x) = 0$, we get $a_1 = 0$. Continuing this process with higher derivatives shows that $a_i = 0$.

Exercise 5.15 *Use the second derivative of the polynomial p in the above example to prove that $a_2 = 0$.*

There is a way to check that a set of sufficiently differentiable real functions is linearly independent, which uses something called the Wronskian. Suppose $f_1, \ldots, f_k$ is a set of functions in the vector space of all C^∞ real functions, which consists of all infinitely differentiable functions. Then we may take derivatives of these functions to obtain functions $f_1', \ldots, f_k'$, then second derivatives $f_1^{(2)}, \ldots, f_k^{(2)}$ and so on until we have the $(k-1)^{st}$ derivatives

$f_1^{(k-1)}, \ldots, f^{(k-1)}$. We can now formally put these functions in as rows of a $(k \times k)$-matrix and take its determinant.

$$W(x) = \det \begin{bmatrix} f_1(x) & f_2(x) & \cdots & f_k(x) \\ f_1'(x) & f_2'(x) & \cdots & f_k'(x) \\ \vdots & \vdots & \vdots & \vdots \\ f_1^{(k-1)}(x) & f_2^{(k-1)}(x) & \cdots & f_k^{(k-1)}(x) \end{bmatrix}$$

This notation is perhaps best understood considering two specific functions, say $f(x) = x^2$ and $g(x) = \sin x$. Since there are two functions, we need to take one derivative, and the Wronskian function is

$$W(x) = \det \begin{bmatrix} x^2 & \sin x \\ 2x & \cos x \end{bmatrix} = x^2 \cos x - 2x \sin x$$

Wroński's Theorem says that a set of functions $f_1, \ldots, f_x$ is linearly independent if and only if the associated Wronskian function is not equal to the zero function. In the above problem, $W(x)$ is clearly not the 0-function, but in order to verify concretely we need only find some x so that $W(x) \neq 0$. For example, $W(\frac{\pi}{2}) = -\pi$. This theorem is a staple of elementary Linear Algebra texts, and therefore I'm mentioning it here. There are important applications of the Wronskian involving linear independence of solution sets of differential equations, in which you may not know exactly what the functions are, but you have information about their derivatives. But the typical exercises in a linear algebra course are no more interesting than the above example because most basic elementary functions learned in calculus, such as trigonometric, exponential, and logarithmic functions are linearly dependent. On the other hand, most math, physics and engineering majors need to learn about differential equations, and may have uses for the Wronskian. Just to give you a little practice, the next exercise will give an alternative argument for linear independence of our standard basis for polynomials. Otherwise, the Wronskian will be of no further use in this text.

Exercise 5.16 *Consider the polynomials $p_0(x) = 1$, $p_1(x) = x$, $p_2(x) = x^2$, $p_3(x) = x^3$.*

> *1. Use the Wronskian to show that these functions are linearly independent by showing that the Wronskian is not the zero function. Hint: When creating the Wronskian matrix, put the functions in order, from the top down, from p_0 to p_3.*

> *2. Generalize the above problem by explaining what will happen when you consider $p_0, p_1, \ldots, p_k$ for any k.*

We will now use *any* basis to define coordinates for any vector, which generalizes the idea of "standard coordinates", i.e. the coordinates $(v_1, \ldots, v_n)$ of a vector in $\mathbb{R}^n$.

Definition 5.4 *If $B = \{\mathbf{v}_1, \ldots, \mathbf{v}_n\}$ is a basis for a vector space V and*

$$\mathbf{v} = \sum_{i=1}^{n} k_i \mathbf{v}_i$$

is the unique expression of $\mathbf{v}$ as a linear combination of the vectors in B, then the scalars k_i are called the coordinates *of $\mathbf{v}$ with respect to the basis B.*

You may have encountered coordinate changes in calculus and/or physics classes. For example, points in the plane can be described using polar coordinates and points in $\mathbb{R}^3$ can be described using spherical and cylindrical coordinates. If you didn't see these, or don't fully remember them, then from the standpoint of this course that doesn't matter because those coordinates are *not* coordinates with respect to a basis–which is the only kind of coordinates that we will consider in this course. But the one of the main reasons to use polar, cylindrical or spherical coordinates is very similar to one of the main reasons to use coordinates with respect to one basis or another: sometimes using different coordinates makes some calculations much simpler, including facilitating and illuminating real-world applications.

We can now interpret the calculation (4.28) as meaning that the "coordinates with respect to the standard basis" are just the "standard coordinates". For $\mathbb{M}_{2\times 2}$, the "standard basis" is

$$M_1 = \begin{bmatrix} 1 & 0 \\ 0 & 0 \end{bmatrix}, \ M_2 = \begin{bmatrix} 0 & 1 \\ 0 & 0 \end{bmatrix}, \ M_3 = \begin{bmatrix} 0 & 0 \\ 1 & 0 \end{bmatrix}, \ M_4 = \begin{bmatrix} 0 & 0 \\ 0 & 1 \end{bmatrix}$$

If this looks a lot like $\mathbb{R}^4$ "rearranged" that's because basically that's what it is. In fact, as we will verify later, but as you already might suspect, all vector spaces having a basis with the same finite number of vectors are "the same" from a purely algebraic standpoint. But let's go ahead and check directly that the standard basis is linearly independent. Suppose that we write the $\mathbf{0}$ matrix as a linear combination of the vectors M_i. That is,

$$\begin{bmatrix} 0 & 0 \\ 0 & 0 \end{bmatrix} = a \begin{bmatrix} 1 & 0 \\ 0 & 0 \end{bmatrix} + b \begin{bmatrix} 0 & 1 \\ 0 & 0 \end{bmatrix} + c \begin{bmatrix} 0 & 0 \\ 1 & 0 \end{bmatrix} + d \begin{bmatrix} 0 & 0 \\ 0 & 1 \end{bmatrix}$$

Simplifying the right side gives us

$$\begin{bmatrix} 0 & 0 \\ 0 & 0 \end{bmatrix} = \begin{bmatrix} a & b \\ c & d \end{bmatrix}$$

which says $a = b = c = d$, as required.

Exercise 5.17 *Show that the standard basis for $\mathbb{M}_{2\times 2}$ does span it. That is, write any matrix $\begin{bmatrix} a & b \\ c & d \end{bmatrix}$ as a linear combination of the matrices M_i, showing that the coordinates of a matrix with respect to the standard basis are simply the entries of the matrix.*

Example 5.13 *Consider the vector space $\mathcal{P}(n)$ of polynomials of degree $\leq n$ which has the standard basis $p_0(x) = 1$, $p_1(x) = x$, ..., $p_n(x) = x^n$. Explicitly, the coordinates of a polynomial $p(x) = a_n x^n + a_{n-1} x^{n-1} + \cdots + a_1 x + a_0$ with respect to the standard basis are just the coefficients: $(a_n, a_{n-1}, \ldots, a_1, a_0)$.*

Determining whether a set of vectors $S = \{\mathbf{v}_1, \ldots, \mathbf{v}_r\}$ in $\mathbb{R}^n$ is a basis requires, according to the definition, showing that S spans $\mathbb{R}^n$ and that S is linearly independent. You have checked linear independence and determined whether individual vectors were in the span of others in previous MPs. Even with the MPs doing the calculations, this was a lot of work—including setting up systems of linear equations and solving them. Therefore you should appreciate how much the next theorem simplifies your work.

Theorem 5.12 *A set $S = \{\mathbf{v}_1, \ldots, \mathbf{v}_r\}$ is a basis for $\mathbb{R}^n$ if and only if $r = n$ and $\det \mathbf{C}_S \neq 0$.*

Proof. If $\mathbf{C}_S$ is not square, the homogeneous system $\mathbf{C}_S \mathbf{X} = \mathbf{0}$ is overdetermined or underdetermined. If $\mathbf{C}_S \mathbf{X} = \mathbf{0}$ is underdetermined, then by Theorem 2.9 the system has infinitely many solutions, and S cannot be linearly independent by Theorem 5.9. On the other hand, if $\mathbf{C}_S \mathbf{X} = \mathbf{0}$ is overdetermined, then Theorem 2.9 tells us that there is a system $\mathbf{C}_S \mathbf{X} = \mathbf{W}$ with no solution. But this means that the vector $\mathbf{w}$ is not in the span of S. Therefore, in order for S to be a basis, $\mathbf{C}_S$ must be square. But if $\det \mathbf{C}_S \neq 0$, then again the system $\mathbf{C}_S \mathbf{X} = \mathbf{0}$ must have infinitely many solutions, so S could not be a basis.

Conversely, suppose that $r = n$ and $\det \mathbf{C}_S \neq 0$. Then any system $\mathbf{C}_S \mathbf{X} = \mathbf{W}$ has a unique solution. But in light of Theorem 5.10, this is just another way of saying that every $\mathbf{w}$ in $\mathbb{R}^n$ can be written uniquely as a linear combination of the vectors in S, meaning that S is a basis. $\blacksquare$

There are two consequences of this theorem that can be quickly derived from Theorem 5.12:

Theorem 5.13 *If $S = \{\mathbf{v}_1, \ldots, \mathbf{v}_r\}$ is a linearly independent set of vectors in $\mathbb{R}^n$ then $r \leq n$.*

Theorem 5.14 *If $S = \{\mathbf{v}_1, \ldots, \mathbf{v}_r\}$ spans $\mathbb{R}^n$ then $r \geq n$.*

Exercise 5.18 *Prove Theorems 5.13 and 5.14.*

We can once again update our list of what it means for a square matrix to be non-singular:

Theorem 5.15 *The following are equivalent for a square matrix $\mathbf{A}$:*

1. $\mathbf{A}$ *is invertible.*

2. *Any equation $\mathbf{AX} = \mathbf{B}$ has a unique solution.*

3. *The homogeneous equation $\mathbf{AX} = \mathbf{0}$ has a unique solution.*

4. *The reduced echelon form of* $\mathbf{A}$ *is* $\mathbf{I}$.

5. $\mathbf{A}$ *is a product of elementary matrices.*

6. $\det \mathbf{A} \neq 0$

7. *The columns of* $\mathbf{A}$ *are a basis for* $\mathbb{R}^n$.

8. *The rows of* $\mathbf{A}$ *are a basis for* $\mathbb{R}^n$.

Proof. The first six properties are equivalent by Theorem 3.10. Parts 6 and 7 are equivalent by Theorem 5.12. ∎

Exercise 5.19 *Use the transpose to prove that Parts 7 and 8 of Theorem 5.15.*

6

Finite Dimensional Vector Spaces

The first goal of this chapter is to define what it means for a vector space to be finite dimensional. We will then show that all finite dimensional vector spaces are "essentially the same" from an algebraic standpoint. This will mean that any general algebraic theorems proved for $\mathbb{R}^n$ can be used with any finite dimensional vector space. We will prove multilinearity of the determinant, describe the Gram-Schmidt orthogonalization process and finish by describing the four fundamental subspaces determined by a linear transformation.

6.1   Dimension

Suppose that $T : V \to W$ is a linear transformation between vector spaces. There are two important properties that T (or for that matter any function) may have. First, T can be *onto*, which means that the range of T is W. Equivalently, for every $\mathbf{w}$ in W there is some $\mathbf{v}$ in V such that $T(\mathbf{v}) = \mathbf{w}$. Another possibility is that T is *one-to-one* (or *1-1*). By definition this means that if $\mathbf{u} \neq \mathbf{v}$ in V, then $T(\mathbf{u}) \neq T(\mathbf{v})$. Put another way, T never takes two different vectors to the same vector. If T is both 1-1 and onto, T is called a *linear isomorphism*. The term "isomorphism" comes from ancient Greek. That is, the prefix "iso" means "the same" and "morph" means "form". So you can think of a linear isomorphism as a function that changes one vector space into another without changing how addition and scalar multiplication operate between corresponding vectors. The term "isomorphism" is used in many contexts in mathematics, but in this course it will *only* mean a *linear isomorphism*.

As an illustration of these concepts we will show that the vector space $\mathbb{M}_{2\times2}$ of 2×2 matrices is isomorphic to $\mathbb{R}^4$. To do so, we will define an isomorphism $T : \mathbb{M}_{2\times2} \to \mathbb{R}^4$ in the completely obvious manner:

$$T\left(\begin{bmatrix} a & b \\ c & d \end{bmatrix}\right) = (a, b, c, d)$$

DOI: 10.1201/9781003674368-6

Let's first check linearity:

$$T\left(k\begin{bmatrix} a & b \\ c & d \end{bmatrix} + \begin{bmatrix} e & f \\ g & h \end{bmatrix}\right) = T\left(\begin{bmatrix} ka + e & kb + f \\ kc + g & kd + h \end{bmatrix}\right)$$

$$= (ka + e, kb + f, kc + g, kd + h) = (ka, kb, kc, kd) + (e, f, g, h)$$

$$= k(a, b, c, d) + (e, f, g, h) = kT\left(\begin{bmatrix} a & b \\ c & d \end{bmatrix}\right) + T\left(\begin{bmatrix} e & f \\ g & h \end{bmatrix}\right)$$

Now let's check that T is 1-1. If $\begin{bmatrix} a & b \\ c & d \end{bmatrix} \neq \begin{bmatrix} e & f \\ g & h \end{bmatrix}$, then these two matrices differ in at least one entry. For example, maybe $b \neq f$. Now

$$T\left(\begin{bmatrix} a & b \\ c & d \end{bmatrix}\right) = (a, b, c, d) \text{ and } T\left(\begin{bmatrix} e & f \\ g & h \end{bmatrix}\right) = (e, f, g, h)$$

and since $b \neq f$, $(a, b, c, d) \neq (e, f, g, h)$.

What about onto? Suppose we have a vector (x, y, z, w) in $\mathbb{R}^4$. Then the matrix $\begin{bmatrix} x & y \\ z & w \end{bmatrix}$ satisfies

$$T\left(\begin{bmatrix} x & y \\ z & w \end{bmatrix}\right) = (x, y, z, w)$$

At any rate these calculations verify what was probably your own sense that 2×2 matrices "are added and multiplied by scalars in the same way that vectors in $\mathbb{R}^4$ are".

Exercise 6.1 *Define $F(a, b, c, d) := \begin{bmatrix} d & b \\ a & c \end{bmatrix}$. Show that F is a linear isomorphism from $\mathbb{R}^4$ to $\mathbb{M}_{2 \times 2}$.*

Exercise 6.2 *I won't ask you to verify that the function $T(a_n x^n + \cdots + a_1 x + a_0) = (a_n, \ldots, a_0)$ is an isomorphism between $\mathcal{P}(n)$ and $\mathbb{R}^{n+1}$—although if you are still unsure about the previous example and exercise then it would be good practice. Instead I'll ask you to verify that the isomorphism T takes the standard basis of $\mathcal{P}(n)$ to the standard basis of $\mathbb{R}^{n+1}$. For example, what is $T(x^{n-1})$?*

The next theorem provides a bit of a shortcut to showing that a linear transformation is 1-1, namely to show that $\mathbf{0}$ is the only vector in $\ker T$.

Theorem 6.1 *If $T : V \to W$ is a linear transformation between vector spaces, T is 1-1 if and only if $\ker T = \{\mathbf{0}\}$.*

Proof. The exact same argument (just after Definition 4.2) as in the case of Euclidean vector spaces shows that $\ker T$ always contains $\mathbf{0}$. If T is 1-1 then by definition, nothing else can be in $\ker T$. This proves that if T is 1-1 then $\ker T = \{\mathbf{0}\}$.

We need to prove the converse statement, namely if $\ker T = \{\mathbf{0}\}$ then T is 1-1. We will prove this by contradiction. Suppose T is *not* 1-1; so there are $\mathbf{v} \neq \mathbf{w}$ such that $T(\mathbf{v}) = T(\mathbf{w})$. Then $\mathbf{v} - \mathbf{w} \neq \mathbf{0}$, but

$$T(\mathbf{v} - \mathbf{w}) = T(\mathbf{v}) - T(\mathbf{w}) = \mathbf{0}.$$

That is, $\ker T$ contains $\mathbf{v} - \mathbf{w}$, which is not $\mathbf{0}$, a contradiction. ∎

The smallest possible vector space consists of only the 0-vector $\{\mathbf{0}\}$. This simple vector space is more useful than it may seem at first. But an important observation is that $\{\mathbf{0}\}$ cannot have a basis. After all, the only possible basis would be $S = \{\mathbf{0}\}$. But we have already seen that any set of vectors that contains $\mathbf{0}$ is not linearly independent. For this reason we treat $\{\mathbf{0}\}$ separately when we define dimension.

Definition 6.1 *The vector space* $\{\mathbf{0}\}$ *is defined to have dimension* 0. *We say that a non-zero vector space V has dimension n if V has a basis $B = \{\mathbf{v}_1, \ldots, \mathbf{v}_n\}$.*

If a non-zero vector space V does not have a finite basis, then V is said to be *infinite dimensional*. There is an important question now: is it possible that V has finite bases having different numbers of vectors, so it has "two different dimensions"? We already know from Theorem 5.12 that every basis of $\mathbb{R}^n$ must have n vectors, and we will use this fact, together with the next important theorem to see that the same statement is true for any finite dimensional vector space.

Theorem 6.2 *If V is a vector space with a finite basis $B = \{\mathbf{v}_1, \ldots, \mathbf{v}_n\}$ then the function $P_B : V \to \mathbb{R}^n$ defined by $P_B\left(\sum_{i=1}^{n} v_i \mathbf{v}_i\right) = (v_1, \ldots, v_n)$ is a linear isomorphism. In particular, every finite dimensional vector space is isomorphic to some $\mathbb{R}^n$.*

Proof. Since B is a basis, according to Theorem 5.11, every $\mathbf{v}$ in V has unique coordinates with respect to B. That is, there is only one way to write $\mathbf{v} = \sum_{i=1}^{n} v_i \mathbf{v}_i$. Therefore the function P_B is well-defined: we have no choice for the numbers v_i. Let's check that P_B is linear. Given a scalar k and vectors $\mathbf{v}$ and $\mathbf{w}$ in V, we may uniquely write $\mathbf{w} = \sum_{i=1}^{n} w_i \mathbf{v}_i$ and calculate:

$$P_B(k\mathbf{v} + \mathbf{w}) = P_B\left(\sum_{i=1}^{n} kk_i\mathbf{v}_i + \sum_{i=1}^{n} w_i\mathbf{v}_i\right) = P_B\left(\sum_{i=1}^{n} (kv_i + w_i)\,\mathbf{v}_i\right)$$

$$= (kv_1 + w_1, \ldots, kv_n + w_n = k(v_1, \ldots, v_n) + (w_1, \ldots, w_n)$$

$$= k\left(P_B \sum_{i=1}^{n} v_i\mathbf{v}_i\right) + P_B\left(\sum_{i=1}^{n} w_i\mathbf{v}_i\right) = kP_B(\mathbf{v}) + P_B(\mathbf{w})$$

Let's check that P_B is one-to one using Theorem 6.1. Suppose that $P_B(\mathbf{v}) = \mathbf{0}$, where $\mathbf{v} = \sum_{i=1}^{n} v_i \mathbf{v_i}$. By definition this means $(v_1, \ldots, v_n) = 0 = (0, \ldots, 0)$ in $\mathbb{R}^n$, which means that $v_i = 0$ for all i and therefore $\mathbf{v} = 0$. The proof of onto is an exercise. ■

Exercise 6.3 *Finish the proof of Theorem 6.2 by showing that the transformation P_B is onto. Hint: For any vector $(w_1, \ldots, w_n)$ in $\mathbb{R}^n$, you need to find a linear combination of the vectors in S that P_B takes to $(w_1, \ldots, w_n)$. This is not as hard as it may seem at first.*

Theorem 6.3 *If $T : V \to W$ is an isomorphism and $B = \{\mathbf{v}_1, \ldots, \mathbf{v}_n\}$ is a basis of V then $B' = \{T(\mathbf{v}_1), \ldots, T(\mathbf{v}_n)\}$ is a basis of W. In particular, if V is finite dimensional then so is W, and the two spaces have the same dimension.*

Proof. Let's check first that B' is linearly independent. Suppose that $\mathbf{0} = \sum_{i=1}^{n} k_i T(\mathbf{v}_i)$; we need to show that $k_i = 0$ for all i. By linearity,

$$T\left(\sum_{i=1}^{n} k_i \mathbf{v}_i\right) = \sum_{i=1}^{n} k_i T(\mathbf{v}_i) = \mathbf{0}$$

That is, $\sum_{i=1}^{n} k_i \mathbf{v}_i$ is in the kernel of T. Since T is one-to one, the only possible vector in $\ker T$ is $\mathbf{0}$, so $\sum_{i=1}^{n} k_i \mathbf{v}_i = \mathbf{0}$. But since B is a basis, this means that $k_i = 0$ for all i.

To show that B' spans W we need to write any $\mathbf{w}$ in W as a linear combination of vectors $T(\mathbf{v}_i)$ in B'. Since T is onto there is some $\mathbf{u} = \sum_{i=1}^{n} u_i \mathbf{v}_i$ such that $T(\mathbf{u}) = \mathbf{w}$. But then by the linearity of T

$$\mathbf{w} = T(\mathbf{u}) = T\left(\sum_{i=1}^{n} u_i \mathbf{v}_i\right) = \sum_{i=1}^{n} u_i T(\mathbf{v}_i).$$

■

We can now check directly that if V is a vector space with a basis $B = \{\mathbf{v}_1, \ldots, \mathbf{v}_n\}$ then every basis of V has n vectors. In fact, suppose that $\{\mathbf{w}_1, \ldots, \mathbf{w}_r\}$ is another basis for V. We already know that $P_B : V \to \mathbb{R}^n$ is an isomorphism, which we have just seen takes bases to bases. This means that $\{P_B(\mathbf{w}_1), \ldots, P_B(\mathbf{w}_r)\}$ would have to be a basis of $\mathbb{R}^n$, and we know already that this means $r = n$. Since every basis of a finite dimensional vector space must have the same size, this shows that no finite dimensional vector space can have "two different dimensions".

Remark 6.1 *Isomorphisms, including P_B, "preserve everything algebraic". One consequence of Theorem 6.3 is that we can use theorems about $\mathbb{R}^n$ and finite dimensional vector spaces interchangeably.*

There is another question, the answer to which might seem obvious at first. If V is a finite dimensional vector space of dimension n and W is a subspace of

V, then is W finite dimensional, and if so is $\dim W \leq n$? If the answer seems to be "obviously yes" then you really need to think about how you would prove such an "obvious" statement. The strategy involves the next theorem, which is important for other reasons.

Theorem 6.4 *(Basis Extension Theorem) Suppose that $S = \{\mathbf{w}_1, \ldots, \mathbf{w}_r\}$ is a linearly independent set of vectors in a vector space. If $\mathbf{v}$ is any vector that is not in the span of S, then $\{\mathbf{w}_1, \ldots, \mathbf{w}_r, \mathbf{v}\}$ is linearly independent.*

You will see how to prove this theorem, as well as how to concretely put it into practice, in the next MP. You will be given a set S of two or three vectors, will build the matrix $\mathbf{C}_S$ and reduce it to a matrix $\mathbf{R}$ in reduced echelon form. At that point you can observe that the corresponding homogeneous system $\mathbf{RX} = \mathbf{0}$ has a unique solution, which verifies that the original vectors are linearly independent, according to Theorem 5.9. You will then augment a vector $\mathbf{B}$ so that the system $\mathbf{RX} = \mathbf{B}$ has no solution. This indicates $\mathbf{B}$ is not in the span of S, according to Theorem 5.10.

MP 61 *BASIS EXTENSION*

Let's turn your experience in the prior MP into a formal proof of Theorem 6.4. Suppose that $\mathbf{v}$ is not in *span* S. To check that $\{\mathbf{w}_1, \ldots, \mathbf{w}_r, \mathbf{v}\}$ is linearly independent, suppose we have a linear combination $\sum_{i=1}^{r} k_i \mathbf{w}_i + k\mathbf{v} = \mathbf{0}$. We need to check that every $k_i = 0$ and $k = 0$. If $k = 0$, we can write

$$\mathbf{0} = \sum_{i=1}^{r} k_i \mathbf{w}_i + k\mathbf{v} = \sum_{i=1}^{r} k_i \mathbf{w}_i$$

Since S is linearly independent, $k_i = 0$ for all i. If $k \neq 0$, then we can write $\mathbf{v} = \sum_{i=1}^{r} \frac{k_i}{k} \mathbf{w}_i$, which is also impossible because $\mathbf{v}$ assumed not to be in the span of S.

Why do we call Theorem 6.4 the Basis Extension Theorem when $S = \{\mathbf{w}_1, \ldots, \mathbf{w}_r\}$ is assumed *not* to *span* V, so it cannot be a basis of V? The reason is that S, being linearly independent, is a basis of *span* S, which is a subspace of V by Theorem 5.6. So we are extending the basis $\{\mathbf{w}_1, \ldots, \mathbf{w}_r\}$ of *span* S to a larger linearly independent set in V, which spans an $(r + 1)$-dimensional subspace. We can continue this process as long as the extended basis does not span V. It is possible that this process never ends—something we will consider below. But if the process does end, then the final linearly independent set spans V, and we have a basis of V. Basis extension has an immediate application in answering our prior question about whether a subspace of a finite dimensional vector space is also finite dimensional. In fact we can actually go a bit farther with this argument:

Theorem 6.5 *Suppose that V is a finite dimensional vector space of dimension n and $W \neq \{0\}$ is a subspace of V. Then W is finite dimensional with dimension $m \leq n$. Moreover, if $W \neq V$ then any linearly independent set $\{\mathbf{w}_1, \ldots, \mathbf{w}_r\}$ in W can be extended to a basis $\{\mathbf{w}_1, \ldots, \mathbf{w}_r, \mathbf{w}_{r+1}, \ldots, \mathbf{w}_n\}$ of V, and therefore $\dim W < \dim V$.*

Proof. To see that W is finite dimensional, start with any non-zero vector $\mathbf{w}_1$ in W. If $\{\mathbf{w}_1\}$ spans W then it is a basis and $\dim W = 1$. If not, then by the Basis Extension Theorem we can extend to a new linearly independent set $\{\mathbf{w}_1, \mathbf{w}_2\}$ of vectors in W. If $\{\mathbf{w}_1, \mathbf{w}_2\}$ spans W, then $\{\mathbf{w}_1, \mathbf{w}_2\}$ is a basis, and $\dim W = 2$. We may continue this extension process, adding new vectors from W. If at any point $\{\mathbf{w}_1, \ldots, \mathbf{w}_m\}$ spans W, then $\dim W = m$. But this process must stop at least when we have n linearly independent vectors in W, because if $S = \{\mathbf{w}_1, \ldots, \mathbf{w}_n\}$ is linearly independent with all $\mathbf{w}_i$ in W, then S must be a basis of V and must span V. But since W is contained in V, S must also span W, and so is a basis of W.

To finish the proof, now that we know W is finite dimensional, we can start the extension process with a basis $\{\mathbf{w}_1, \ldots, \mathbf{w}_r\}$ of W. If $W \neq V$ then we can extend to a linearly independent set $\{\mathbf{w}_1, \ldots, \mathbf{w}_r, \mathbf{v}_{r+1}\}$ using some $\mathbf{v}_{r+1}$ not in W. If $r + 1 = n$, then this new set is a basis of V and $\dim V = n = r + 1 > r = \dim W$, and we are finished. If not, then we just continue this process until we reach a linearly independent set $\{\mathbf{w}_1, \ldots, \mathbf{w}_r, \mathbf{v}_{r+1}, \ldots, \mathbf{v}_n\}$ of V, which must be a basis for V–and by construction $\dim V = n > r = \dim W$. ∎

Now suppose that we don't know *a priori* (from the beginning) that V is finite dimensional, and start the same process with some non-zero vector in V. If it spans V then this proves that V is one-dimensional. If it does not span V, then we can extend to two linearly independent vectors, and so on. The above theorem tells us that if V is finite dimensional then this process must end. Therefore, if the process does not end, V must be infinite dimensional. For example, consider the space $\mathcal{P}$ of all polynomials (with no restriction on the degree). We have already seen that $1, x, x^2, \ldots, x^n$ are linearly independent. That is $\mathcal{P}$ contains linearly independent sets of all possible sizes, so by the above theorem $\mathcal{P}$ must be infinite dimensional. Another very important infinite dimensional vector space is the set $\mathbb{R}^\infty$ of all infinite sequences $\mathbf{x} = (x_1, \ldots, x_n, \ldots)$ of real numbers, as mentioned previously.

We conclude this section with an observation that is sometimes useful. If a finite set $S = \{\mathbf{w}_1, \ldots, \mathbf{w}_n\}$ spans a non-zero vector space V but is not a basis, then it is not linearly independent. According to Theorem 5.7, we can remove one element of S to obtain a strictly smaller set S', which still spans V. If S' is linearly dependent, we can remove another element to leave a strictly smaller set S'' that still spans V. Since V is non-zero, this process must eventually stop with a linearly independent set that spans V, which is therefore a basis for V. We have proved the following theorem, which is a kind of counterpart to the Basis Extension Theorem:

Theorem 6.6 *(Spanning Set Reduction) If $S = \{\mathbf{w}_1, \ldots, \mathbf{w}_n\}$ is a subset of a vector space V that spans V, then some subset of S is a basis for V.*

Remark 6.2 *I would like to draw your attention to one important distinction between Basis Extension and Spanning Set Reduction. When extending a linearly independent set that does not span a vector space, you are free to choose any vector that is not in the span of that linearly independent set, and adding that vector will produce a larger linearly independent set. However, when reducing a spanning set that is not linearly independent, you are **not** free to simply toss out any vector from the set. Doing so might reduce the size of the span of the vectors. You must find a vector that is a linear combination of the others, and remove that "redundant" one.*

Exercise 6.4 *Find three vectors $\{\mathbf{v}_1, \mathbf{v}_2, \mathbf{v}_3\}$ in $\mathbb{R}^3$ such that removing $\mathbf{v}_1$ leaves a linearly dependent set but removing $\mathbf{v}^3$ leaves a linearly independent set.*

6.2 Multilinearity of the Determinant

Multilinearity is not always included in elementary linear algebra courses; when it is, it is often referred to as "row additivity". This section verifies multilinearity of the determinant, uses multilinearity to show that cofactor expansion computes the derivative, and shows that the determinant function is also uniquely characterized by its multilinearity property.

While the determinant of the product of two matrices is always equal to the product of their determinants, the same is not true for sums–even for so simple a function as the identity function! For example

$$\det\left(\begin{bmatrix} 1 & 0 \\ 0 & 1 \end{bmatrix} + \begin{bmatrix} 1 & 0 \\ 0 & 1 \end{bmatrix}\right) = \det\begin{bmatrix} 2 & 0 \\ 0 & 2 \end{bmatrix} = 4$$

but

$$\det\begin{bmatrix} 1 & 0 \\ 0 & 1 \end{bmatrix} + \det\begin{bmatrix} 1 & 0 \\ 0 & 1 \end{bmatrix} = 1 + 1 = 2$$

However, the determinant does have a very important property involving sums, called *multilinearity* that we will define below. This property may seem complicated at first, but it is extremely important. Multilinear functions occur throughout science and engineering because, again, they can be used to describe real phenomena in nature. For example, physics and many kinds of engineering use functions called *tensors*, which are multilinear operators, and every student of these subjects, as well as mathematics students, should understand multilinearity.

Recall that we have denoted the ith row of a matrix $\mathbf{A}$ by $\mathbf{A}_i$ and the jth column by $\mathbf{A}^j$. For a square matrix

$$\mathbf{A} = \begin{bmatrix} a_{11} & a_{12} & \cdots & a_{1n} \\ a_{21} & a_{22} & \cdots & a_{2n} \\ \vdots & \vdots & \vdots & \vdots \\ a_{n1} & a_{n2} & \cdots & a_{nn} \end{bmatrix},$$

$\mathbf{A}_i = \begin{bmatrix} a_{i1} & a_{i2} & \cdots & a_{in} \end{bmatrix}$ and we can write $\mathbf{A}$ as a block column of its rows:

$$\{\mathbf{A} = \begin{bmatrix} \mathbf{A}_1 \\ \mathbf{A}_2 \\ \vdots \\ \mathbf{A}_n \end{bmatrix}$$

Now suppose that $\mathbf{R} = \begin{bmatrix} r_1 & r_2 & \cdots & r_n \end{bmatrix}$ is a row vector and k is a scalar. We will create a new matrix as follows:

$$\begin{bmatrix} \mathbf{A}_1 \\ \vdots \\ \mathbf{A}_i + k\mathbf{R} \\ \vdots \\ \mathbf{A}_n \end{bmatrix}$$

That is, we are adding the scalar multiple of the row $\mathbf{R}$ to the ith row of $\mathbf{A}$. This is the same as the row operation RO2, except that $\mathbf{R}$ does *not* have to be a row in $\mathbf{A}$.

Multilinearity says that

$$\det \begin{bmatrix} \mathbf{A}_1 \\ \vdots \\ \mathbf{A}_i + k\mathbf{R} \\ \vdots \\ \mathbf{A}_n \end{bmatrix} = \det \begin{bmatrix} \mathbf{A}_1 \\ \vdots \\ \mathbf{A}_i \\ \vdots \\ \mathbf{A}_n \end{bmatrix} + k \det \begin{bmatrix} \mathbf{A}_1 \\ \vdots \\ \mathbf{R} \\ \vdots \\ \mathbf{A}_n \end{bmatrix} \qquad (6.1)$$

Note that the middle term is just $\det \mathbf{A}$, but we have written it in this way to better illustrate multilinearity. This formula should remind you of the calculation that you did in Exercise 3.1, in which you characterized linear real functions as being precisely those functions f such that $f(a + kb) = f(a) + kf(b)$. Visually speaking, you are "splitting f across the sum and pulling out the scalar k". In this case, linearity takes place in a row, and the determinant function is being considered as a function of those rows.

To simplify our notation we will prove multilinearity in the top row; the proofs for the other rows are basically the same. Consider two cases: $\mathbf{R}$ is in the span of $S = \{\mathbf{A}_2, \ldots, \mathbf{A}_n\}$, and $\mathbf{R}$ is not in the span. In the first case, $\mathbf{R} = c_2 \mathbf{A}_2 + \cdots + c_n \mathbf{A}_n$ and therefore the row $\mathbf{A}_1 + k\mathbf{R}$ can be obtained from $\mathbf{A}_1$ by repeatedly adding a scalar multiples of other rows (i.e. using RO2). As you know, RO2 does not change the determinant, so

$$\det \begin{bmatrix} \mathbf{A}_1 + k\mathbf{R} \\ \vdots \\ \mathbf{A}_i \\ \vdots \\ \mathbf{A}_n \end{bmatrix} = \det \begin{bmatrix} \mathbf{A}_1 \\ \vdots \\ \mathbf{A}_i \\ \vdots \\ \mathbf{A}_n \end{bmatrix} = \det \mathbf{A}$$

On the other hand, $\mathbf{R}$ is in the span of the remaining rows, so the matrix

$$\begin{bmatrix} \mathbf{R} \\ \vdots \\ \mathbf{A}_i \\ \vdots \\ \mathbf{A}_n \end{bmatrix}$$

has linearly dependent rows. By Theorem 5.15 this matrix has determinant 0, finishing the proof when $\mathbf{R}$ is in the span of S.

If $\mathbf{R}$ is not in the span of S then we can consider two cases: S is linearly dependent or S linearly independent. In the first case, Theorem 5.15 tells us that all three matrices in the multilinearity Equation (6.1) have determinant 0, and the equation is simply $0 = 0$, which is true. Finally, if S is linearly independent then since $\mathbf{R}$ is not in the span of S, by the Basis Extension Theorem, $\{\mathbf{R}, \mathbf{A}_2, \ldots, \mathbf{A}_n\}$ is a basis. This means that for some scalars $c_1, \ldots, c_n$,

$$\mathbf{A}_1 = c_1 \mathbf{R} + c_2 \mathbf{A}_2 + \cdots + c_n \mathbf{A}_n = c_1 \mathbf{R} + \sum_{i=2}^{n} c_i \mathbf{A}_i \qquad (6.2)$$

We now compute:

$$\det \begin{bmatrix} \mathbf{A}_1 + k\mathbf{R} \\ \vdots \\ \mathbf{A}_i \\ \vdots \\ \mathbf{A}_n \end{bmatrix} = \det \begin{bmatrix} (c_1 + k)\mathbf{R} + \sum_{i=2}^{n} c_i \mathbf{A}_i \\ \vdots \\ \mathbf{A}_i \\ \vdots \\ \mathbf{A}_n \end{bmatrix} = \det \begin{bmatrix} (c_1 + k)\mathbf{R} \\ \vdots \\ \mathbf{A}_i \\ \vdots \\ \mathbf{A}_n \end{bmatrix}$$

$$= (c_1 + k)\det \begin{bmatrix} \mathbf{R} \\ \vdots \\ \mathbf{A}_i \\ \vdots \\ \mathbf{A}_n \end{bmatrix} = k \det \begin{bmatrix} \mathbf{R} \\ \vdots \\ \mathbf{A}_i \\ \vdots \\ \mathbf{A}_n \end{bmatrix} + \det \begin{bmatrix} c_1 \mathbf{R} \\ \vdots \\ \mathbf{A}_i \\ \vdots \\ \mathbf{A}_n \end{bmatrix}$$

$$= k \det \begin{bmatrix} \mathbf{R} \\ \vdots \\ \mathbf{A}_i \\ \vdots \\ \mathbf{A}_n \end{bmatrix} + \det \begin{bmatrix} \mathbf{A}_1 - \sum_{i=2}^{n} c_i \mathbf{A}_i \\ \vdots \\ \mathbf{A}_i \\ \vdots \\ \mathbf{A}_n \end{bmatrix} = k \det \begin{bmatrix} \mathbf{R} \\ \vdots \\ \mathbf{A}_i \\ \vdots \\ \mathbf{A}_n \end{bmatrix} + \det \begin{bmatrix} \mathbf{A}_1 \\ \vdots \\ \mathbf{A}_i \\ \vdots \\ \mathbf{A}_n \end{bmatrix}$$

Exercise 6.5 *For each "=" in the above calculation, state the reasons. Hint: D3 is used twice.*

Multilinearity allows us to finally check that the function D computed via cofactor expansion is equal to the determinant. I'll demonstrate the iterative process going from 2×2 matrices to 3×3 matrices. This proof can be modified to prove the statement for matrices of any size. Let's consider the determinant (defined using Gauss-Jordan elimination) of a matrix of the following form:

$$\begin{bmatrix} 1 & 0 & 0 \\ a_{21} & a_{22} & a_{23} \\ a_{31} & a_{32} & a_{33} \end{bmatrix}$$

Suppose that $a_{22} \neq 0$. Using three row operations that don't change the determinant we can reduce to

$$\begin{bmatrix} 1 & 0 & 0 \\ 0 & a_{22} & a_{23} \\ 0 & a_{32} & a_{33} \end{bmatrix} \rightarrow \begin{bmatrix} 1 & 0 & 0 \\ 0 & a_{22} & a_{23} \\ 0 & 0 & a_{33} - \frac{a_{32}}{a_{22}} a_{23} \end{bmatrix}$$

This matrix is upper triangular, so its determinant is

$$1 \left(a_{22} \left(a_{33} - \frac{a_{32}}{a_{22}} a_{23} \right) \right) = a_{22} a_{33} - a_{32} a_{23}$$

By now you recognize that the determinant of this matrix is equal to the determinant of the $(1,1)$ minor matrix $\mathbf{M}_{11} = \begin{bmatrix} a_{22} & a_{23} \\ a_{32} & a_{33} \end{bmatrix}$. In other words, for a matrix of this form, the determinant calculated by Gauss-Jordan elimination is the same as the determinant calculated by cofactor expansion. The case when $a_{22} = 0$ is an exercise.

Exercise 6.6 *Finish the proof of Multilinearity by arguing the case when $a_{22} = 0$. Hint: First say what happens if a_{32} is also 0, and when this is not the case, swap rows.*

Exercise 6.7 *Show that the determinant of* $\begin{bmatrix} 0 & 1 & 0 \\ a_{21} & a_{22} & a_{23} \\ a_{31} & a_{32} & a_{33} \end{bmatrix}$ *is* $-\det \mathbf{M}_{12}$.
*You may **not** use cofactor expansion in this proof because we're in the midst of proving that cofactor expansion gives us the determinant, and using cofactor expansion would be circular reasoning. Instead, swap the first two columns first, which introduces a minus sign (see Theorem 3.9).*

Likewise, $\det \begin{bmatrix} 0 & 0 & 1 \\ a_{21} & a_{22} & a_{23} \\ a_{31} & a_{32} & a_{33} \end{bmatrix} = \det \mathbf{M}_{13}$ (this time you have to swap columns twice, so the minus signs cancel). Now we can simply apply multilinearity of the determinant:

$$\det \begin{bmatrix} a_{11} & a_{12} & a_{13} \\ a_{21} & a_{22} & a_{23} \\ a_{31} & a_{32} & a_{33} \end{bmatrix}$$

$$= \det \begin{bmatrix} a_{11} \begin{bmatrix} 1 & 0 & 0 \end{bmatrix} + a_{12} \begin{bmatrix} 0 & 1 & 0 \end{bmatrix} + a_{13} \begin{bmatrix} 0 & 0 & 1 \end{bmatrix} \\ \mathbf{A}_2 \\ \mathbf{A}_3 \end{bmatrix}$$

$$= a_{11} \det \begin{bmatrix} 1 & 0 & 0 \\ a_{21} & a_{22} & a_{23} \\ a_{31} & a_{32} & a_{33} \end{bmatrix} + a_{12} \det \begin{bmatrix} 0 & 1 & 0 \\ a_{21} & a_{22} & a_{23} \\ a_{31} & a_{32} & a_{33} \end{bmatrix}$$

$$+ a_{13} \det \begin{bmatrix} 0 & 0 & 1 \\ a_{21} & a_{22} & a_{23} \\ a_{31} & a_{32} & a_{33} \end{bmatrix} = a_{11} \det \mathbf{M}_{11} - a_{12} \det \mathbf{M}_{12} + a_{13} \det \mathbf{M}_{13} = D(\mathbf{A})$$

This shows that Gauss-Jordan elimination and cofactor expansion give the same determinant number for a 3×3 matrix. We can extend this to 4×4 matrices in a similar fashion, by considering the matrices

$$\begin{bmatrix} 1 & 0 & 0 & 0 \\ a_{21} & a_{22} & a_{23} & a_{24} \\ a_{31} & a_{32} & a_{33} & a_{34} \\ a_{41} & a_{42} & a_{43} & a_{44} \end{bmatrix}, \begin{bmatrix} 0 & 1 & 0 & 0 \\ a_{21} & a_{22} & a_{23} & a_{24} \\ a_{31} & a_{32} & a_{33} & a_{34} \\ a_{41} & a_{42} & a_{43} & a_{44} \end{bmatrix} \cdots$$

and using multilinearity. Formally speaking the proof is by induction, a topic that we will touch on in more detail later. We conclude this chapter with an important uniqueness theorem.

Theorem 6.7 *The determinant function is the unique function Δ defined for square matrices that is multilinear in rows, such that $\Delta(\mathbf{I}) = 1$.*

Proof. You have seen this drill before. By uniqueness, we need only show that the multilinear function Δ satisfies D2 and D3. Multiplication of a row $\mathbf{A}_i$ by a scalar c is the same as adding $(c-1)$ times that row to itself. Let's see what happens when we apply multilinearity:

$$\Delta\left(\begin{bmatrix} \mathbf{A}_1 \\ \vdots \\ c\mathbf{A}_i \\ \vdots \\ \mathbf{A}_n \end{bmatrix}\right) = \Delta\left(\begin{bmatrix} \mathbf{A}_1 \\ \vdots \\ \mathbf{A}_i + (c-1)\mathbf{A}_i \\ \vdots \\ \mathbf{A}_n \end{bmatrix}\right)$$

$$= \Delta \left(\begin{bmatrix} \mathbf{A}_1 \\ \vdots \\ \mathbf{A}_i \\ \vdots \\ \mathbf{A}_n \end{bmatrix} \right) + (c-1)\Delta \left(\begin{bmatrix} \mathbf{A}_1 \\ \vdots \\ \mathbf{A}_i \\ \vdots \\ \mathbf{A}_n \end{bmatrix} \right)$$

$$= c\Delta \left(\begin{bmatrix} \mathbf{A}_1 \\ \vdots \\ \mathbf{A}_i \\ \vdots \\ \mathbf{A}_n \end{bmatrix} \right)$$

This proves D2, and the proof of D3 is an exercise. ∎

Exercise 6.8 *Finish the proof of Theorem 6.7 as follows:*

1. Show that if Δ is multilinear and a matrix $\mathbf{A}$ has a row of 0s then $\Delta(\mathbf{A}) = 0$. That is, $\Delta \left(\begin{bmatrix} \mathbf{A}_1 \\ \vdots \\ \mathbf{0} \\ \vdots \\ \mathbf{A}_n \end{bmatrix} \right) = 0$. Hint: use $\mathbf{0} = \mathbf{0} + \mathbf{0}$.

2. Use the first part of this exercise to prove D3. That is, apply multilinearity to show that for $i \neq j$,

$$\Delta \left(\begin{bmatrix} \mathbf{A}_1 \\ \vdots \\ \mathbf{A}_i + k\mathbf{A}_j \\ \vdots \\ \mathbf{A}_j \\ \vdots \\ \mathbf{A}_n \end{bmatrix} \right) = \Delta \left(\begin{bmatrix} \mathbf{A}_1 \\ \vdots \\ \mathbf{A}_i \\ \vdots \\ \mathbf{A}_j \\ \vdots \\ \mathbf{A}_n \end{bmatrix} \right)$$

Exercise 6.9 *Recall the definition of the cross product using the cofactor expansion determinant formula*

$$\mathbf{u} \times \mathbf{v} = \det \begin{bmatrix} \mathbf{i} & \mathbf{j} & \mathbf{k} \\ u_1 & u_2 & u_3 \\ v_1 & v_2 & v_3 \end{bmatrix}$$

Formally apply multilinearity of the determinant to derive the following formula from Section 4.5:

$$\mathbf{u} \times (k\mathbf{v} + \mathbf{w}) = k(\mathbf{u} \times \mathbf{v}) + \mathbf{u} \times \mathbf{w}$$

MP 62 *MULTILINEARITY*

6.3 Geometric Interpretation of Dimension and Subspaces

Because all n-dimensional vectors spaces are isomorphic to $\mathbb{R}^n$, there is no loss in generality in confining our geometric discussion to $\mathbb{R}^n$ and using your geometric intuition. $\mathbb{R}^0$ is simply $\{\mathbf{0}\}$. Intuitively speaking a point has no dimension, so it is appropriate that we say $\mathbb{R}^0$ is 0-dimensional. $\mathbb{R}^1$ just consists of 1-vectors $[v]$, so $\mathbb{R}^1$ is geometrically the real line $\mathbb{R}$. Generally we simply use the notation $\mathbb{R}$ rather than $\mathbb{R}^1$. $\mathbb{R}^2$ is just the plane. But we now have the possibility of non-zero proper subspaces of $\mathbb{R}^2$. Such a subspace W must be of dimension 1, and so has a basis consisting of a single non-zero vector $\mathbf{v}$. The span W of $\mathbf{v}$ consists of all linear combinations in the set $S = \{\mathbf{v}\}$. Since there is only one vector in S, $W = \{t\mathbf{v}\}$ for all real numbers t. That is, as you have seen already, W is geometrically a line through the origin. In $\mathbb{R}^3$, the one-dimensional subspaces are likewise of the form $W = \{t\mathbf{v}\}$, and geometrically are also lines through the origin. Given any two non-zero vectors $\mathbf{v}$ and $\mathbf{w}$, it is possible that one is a non-zero scalar multiple of the other, for example, $\mathbf{v} = k\mathbf{w}$. This means that the space spanned by them is one-dimensional, and so is a line through the origin. If neither is a scalar multiple of the other then they are linearly independent and span a two-dimensional subspace P. According to Theorem 6.2, P is isomorphic to $\mathbb{R}^2$, and is geometrically a plane through the origin. If we take three vectors $\mathbf{v}_1$, $\mathbf{v}_2$, $\mathbf{v}_3$, then if they are linearly independent, they span $\mathbb{R}^3$. If they are linearly dependent, then they must span a two-dimensional subspace or a one-dimensional subspace, i.e. a line or a plane. If we take four or more non-zero vectors in $\mathbb{R}^3$, then by they must be linearly dependent. Likewise a set of non-zero vectors may span all of $\mathbb{R}^4$ (in which case there must be at least four of them), a three-dimensional subspace isomorphic to $\mathbb{R}^3$ (there must be at least three of them) a two-dimensional subspace isomorphic to $\mathbb{R}^2$ (there must be at least two) or a subspace isomorphic to $\mathbb{R}$. The same kind of geometric interpretations apply to $\mathbb{R}^n$.

6.4 Change of Coordinates

If we choose a basis $B = \{\mathbf{v}_1, \ldots, \mathbf{v}_n\}$ for $\mathbb{R}^n$, then any $\mathbf{w} = (w_1, \ldots, w_n)$ in $\mathbb{R}^n$ can be uniquely expressed as $\mathbf{w} = \sum_{i=1}^{n} k_i \mathbf{v}_i$. The vector $(k_1, \ldots, k_n)$ is the *coordinate expression* of **w** *with respect to the basis B* (or just the *coordinates of* **w** *in B*). Note that by definition, $(k_1, \ldots, k_n) = P_B(\mathbf{w})$ (see Theorem 6.2). From a practical standpoint, given a vector $\mathbf{w} = (w_1, \ldots, w_n)$, how do you calculate $P_B(\mathbf{w})$? You already know how to do this, and the next MP will serve as a reminder.

MP 63 *COORDINATES WITH RESPECT TO A BASIS*

Let's think again about what you just did in the previous MP. Throughout this argument we refer to Theorem 5.15. First you created the augmented matrix $\mathbf{C}_B \mid \mathbf{W}$ and used row operations to solve the system $\mathbf{C}_B \mathbf{X} = \mathbf{W}$ to obtain $\mathbf{X} = P_B(\mathbf{w}) = (k_1, \ldots, k_n)$. You may have noticed that there was always a unique solution–this is because B is a basis so there is a unique expression $\mathbf{w} = \sum_{i=1}^{n} k_i \mathbf{v}_i$. This in turn means that $\mathbf{C}_B$ is invertible and you may solve the system $\mathbf{C}_B \mathbf{X} = \mathbf{W}$ as $\mathbf{X} = \mathbf{C}_B^{-1} \mathbf{W}$.

Now suppose that we have two bases $B = \{\mathbf{v}_1, \ldots, \mathbf{v}_n\}$ and $B' = \{\mathbf{v}_1', \ldots, \mathbf{v}_n'\}$ in $\mathbb{R}^n$. For any vector $\mathbf{w} = (w_1, \ldots, w_n)$ we may write the coordinate expressions $\mathbf{w} = \sum_{i=1}^{n} k_i \mathbf{v}_i$ in B and $\mathbf{w} = \sum_{i=1}^{n} k_i' \mathbf{v}_i'$ in B'. If we are given $(k_1, \ldots, k_n)$, how do we find $(k_1', \ldots, k_n')$? In order to recover the standard coordinates of **w**, we multiply $\mathbf{C}_B \begin{bmatrix} k_1 \\ k_2 \\ \vdots \\ k_n \end{bmatrix} = \begin{bmatrix} w_1 \\ w_2 \\ \vdots \\ w_n \end{bmatrix}$ and then we can multiply

$$\mathbf{C}_{B'}^{-1} \begin{bmatrix} w_1 \\ w_2 \\ \vdots \\ w_n \end{bmatrix} = \begin{bmatrix} k_1' \\ k_2' \\ \vdots \\ k_n' \end{bmatrix}$$

That is,

$$\begin{bmatrix} k_1' \\ k_2' \\ \vdots \\ k_n' \end{bmatrix} = \mathbf{C}_{B'}^{-1} \mathbf{C}_B \begin{bmatrix} k_1 \\ k_2 \\ \vdots \\ k_n \end{bmatrix}$$

In other words, $\mathbf{C}_{B'}^{-1} \mathbf{C}_B$ is the matrix that represents the linear isomorphism that changes from coordinates in B to coordinates in B'.

Definition 6.2 *Given two bases B_1 and B_2 for $\mathbb{R}^n$, we define* $\mathbf{P}_{B_1 B_2} := \mathbf{C}_{B_2}^{-1} \mathbf{C}_{B_1}$ *and call it the transition matrix from the basis B_1 to the basis B_2.*

Note that by definition,

$$\left(\mathbf{P}_{B_1 B_2} \right)^{-1} = \left(\mathbf{C}_{B_2}^{-1} \mathbf{C}_{B_1} \right)^{-1} = \mathbf{C}_{B_1}^{-1} \mathbf{C}_{B_2} = \mathbf{P}_{B_2 B_1} \tag{6.3}$$

The above formula should make perfect sense to you: The inverse of the matrix that changes coordinates from B_1 to B_2 should be the matrix that changes coordinates from B_2 to B_1. One can use Gauss-Jordan elimination to obtain the transition matrix. In fact, if you begin with the augmented matrix $\begin{bmatrix} \mathbf{C}_{B_2} & | & \mathbf{C}_{B_1} \end{bmatrix}$ and reduce the left matrix to the identity, then you are multiplying on the left by elementary matrices whose product is $\mathbf{C}_{B_2}^{-1}$, so the augmented matrix becomes $\begin{bmatrix} \mathbf{I} & | & \mathbf{C}_{B_2}^{-1}\mathbf{C}_{B_1} \end{bmatrix}$. You will practice this in the next MP.

MP 64 *TRANSITION MATRIX*

6.5   Matrix Expression of a Linear Transformation

Since all finite dimensional vector spaces are isomorphic to some Euclidean $\mathbb{R}^n$, one should be able to represent any linear transformation $T : V \to W$ between finite dimensional vector spaces using matrices. To make this precise, choose a basis $B_V = \{\mathbf{v}_1, \ldots, \mathbf{v}_n\}$ for V and a basis $B_W = \{\mathbf{w}_1, \ldots, \mathbf{w}_k\}$ for W. Then for each $\mathbf{v}_i$, we may uniquely write $T(\mathbf{v}_i) = \sum_{j=1}^{k} t_{ji}\mathbf{w}_j$, and we have an $k \times n$ matrix

$$\mathbf{A}_{B_V B_W} = \begin{bmatrix} t_{11} & t_{12} & \cdots & t_{1n} \\ t_{21} & t_{22} & \cdots & t_{2n} \\ \vdots & \vdots & \vdots & \vdots \\ t_{k1} & t_{k2} & \cdots & t_{kn} \end{bmatrix} \tag{6.4}$$

We will call $\mathbf{A}_{B_V B_W}$ the *matrix representation of T with respect to the bases B_V and B_W*. You should notice that the ith column of $\mathbf{A}_{B_V B_W}$ is *precisely* the coordinate expression of the vector $T(\mathbf{v}_i)$ with respect to the basis B_W, which is also written as $P_{B_W}(T(\mathbf{v}_i))$. In other words, if

$$S = \{P_{B_W}(T(\mathbf{v}_1)), \ldots, P_{B_W}(T(\mathbf{v}_n))\} \tag{6.5}$$

then

$$\mathbf{A}_{B_V B_W} = [P_{B_W}(T(\mathbf{v}_1)) \mid \cdots \mid P_{B_W}(T(\mathbf{v}_n))] = \mathbf{C}_S \tag{6.6}$$

Does this process give us the old familiar matrix from Theorem 4.11 when B_V and B_W are the standard bases for $\mathbb{R}^n$ and $\mathbb{R}^k$? In that case the matrix was $\mathbf{C}_S$, where $S = \{T(\mathbf{e}_1), \ldots, T(\mathbf{e}_n)\}$. If we use the standard basis in $\mathbb{R}^n$, we replace $\mathbf{v}_i$ by $\mathbf{e}_i$ in the (6.5). Moreover, if E is the standard basis in $\mathbb{R}^k$ and $\mathbf{v} = (v_1, \ldots, v_k) = \sum_{i=1}^{k} v_i\, \mathbf{e}_i$, then $P_E(\mathbf{v}) = (v_1, \ldots, v_n)$. Therefore $P_E(T(\mathbf{v}_i)) = T(\mathbf{v}_i)$ in (6.5). In other words in the special special case of the standard bases in the domain and codomain, $S = \{T(\mathbf{e}_1), \ldots, T(\mathbf{e}_n)\}$, exactly as in Theorem 4.11.

In the next MP you will practice finding the matrix expression of a linear transformation from $\mathbb{R}^2$ to $\mathbb{R}^2$ with a *non-standard* basis $B = \{\mathbf{u}, \mathbf{v}\}$ for the domain and the standard basis for the codomain, which is $\mathbf{C}_S$, where $S = \{T(\mathbf{u}), T(\mathbf{v})\}$.

MP 65 *EXPRESSION WITH NS DOMAIN BASIS*

In the next MP you will do the same thing for a linear transformation $T_{\mathbf{A}}$, with the standard basis in the domain but a non-standard basis $B = \{\mathbf{u}, \mathbf{v}\}$ in the codomain. With respect to the standard basis in the codomain, the matrix expression of $T_{\mathbf{A}}$ is of course simply $\mathbf{A}$. But you then need to express the columns of $\mathbf{A}$ in terms of the new basis B, which is something that you practiced in MP 63. That is, you need to find the matrix $\mathbf{C}_B^{-1}\mathbf{A}$ because the columns of that matrix are precisely the coordinate expressions of the columns of $\mathbf{A}$ with respect to the new basis.

MP 66 *EXPRESSION WITH NS CODOMAIN BASIS*

Of course the matrix $\mathbf{A}_{B_V B_W}$ depends on the choice of the two bases. But this opens the very real possibility that by choosing one's bases carefully, one might obtain a particularly "nice" matrix representing the transformation. It is important to keep in mind that the original function—the transformation itself—*does not change* when we change basis. The change of basis simply gives us a different matrix with which to compute it. In this way, change of basis is extremely important even for the vector spaces $\mathbb{R}^n$, because the standard basis might not be the "best" basis to represent a given transformation. The representation of linear transformations by matrices also shows that every linear transformation between finite dimensional vector spaces can be understood via matrix algebra, and we will begin to do so in the next section.

Before moving on, however, let's see how basis change impacts the matrix of a transformation $T : \mathbb{R}^n \to \mathbb{R}^n$. In this case, we can consider the domain and codomain as the "same vector space", which is useful for many applications. For example, you have already seen that rotations can be expressed as matrices. Since the domain and codomain are the same, for any basis $B = \{\mathbf{v}_1, \ldots, \mathbf{v}_n\}$ of $\mathbb{R}^n$, it is natural to consider the matrix representation of T with respect to B as the basis for *both* the domain and codomain. That is, we are interested in the matrix $\mathbf{A}_{BB}$, which will simply refer to as $\mathbf{A}_B$.

Let's recap how to use the matrix $\mathbf{A}_B$. First, $\mathbf{C}_B$ is the matrix whose columns are the coordinates with respect to the B of the vectors $T(\mathbf{v}_i)$. Given a vector $\mathbf{w} = (w_1, \ldots, w_n)$ with corresponding column vector $\mathbf{W} = \begin{bmatrix} w_1 \\ \vdots \\ w_n \end{bmatrix}$, in order to use the matrix $\mathbf{A}_B$ to find $T(\mathbf{w})$ we first find the component expression of $\mathbf{w}$ with respect to B, i.e. $\mathbf{C}_B^{-1}\mathbf{W}$. We then multiply on the left by $\mathbf{A}_B$, and then on the left by $\mathbf{C}_B$ to get the coordinates of $T(\mathbf{w})$ with respect to B. We have shown:

Theorem 6.8 *If $T : \mathbb{R}^n \to \mathbb{R}^n$ is a linear and B is a basis for $\mathbb{R}^n$, then for any vector $\mathbf{w} = (w_1, \ldots, w_n)$ in $\mathbb{R}^n$,*

$$T(\mathbf{w}) = \mathbf{C}_B \mathbf{A}_B \mathbf{C}_B^{-1}\mathbf{W}$$

Now suppose we have two bases B_1 and B_2 for $\mathbb{R}^n$. Then we have two matrices $\mathbf{A}_{B_1}$ and $\mathbf{A}_{B_1}$ representing T, and naturally we would like to know what is the relationship between these matrices. We can now write $T(\mathbf{w})$ in two ways for any $\mathbf{w} = (w_1, \ldots, w_n)$:

$$T(\mathbf{w}) = \mathbf{C}_{B_1} \mathbf{A}_{B_1} \mathbf{C}_{B_1}^{-1}\mathbf{W} \text{ and } T(\mathbf{w}) = \mathbf{C}_{B_2} \mathbf{A}_{B_2} \mathbf{C}_{B_2}^{-1}\mathbf{W}$$

Since these equations hold for every $\mathbf{w}$, the uniqueness of Theorem 4.10 tells us that

$$\mathbf{C}_{B_1} \mathbf{A}_{B_1} \mathbf{C}_{B_1}^{-1} = \mathbf{C}_{B_2} \mathbf{A}_{B_2} \mathbf{C}_{B_2}^{-1}$$

Using matrix algebra, the fact that $\mathbf{P}_{B_2 B_1} = \mathbf{C}_{B_1}^{-1}\mathbf{C}_{B_2}$, and Formula (6.3) we obtain that

$$\mathbf{A}_{B_2} = \mathbf{C}_{B_2}^{-1}\mathbf{C}_{B_1} \mathbf{A}_{B_1} \mathbf{C}_{B_1}^{-1}\mathbf{C}_{B_2} = \left(\mathbf{P}_{B_2 B_1}\right)^{-1} \mathbf{A}_{B_1} \left(\mathbf{P}_{B_2 B_1}\right)$$

$$= \left(\mathbf{P}_{B_1 B_2}\right) \mathbf{A}_1 \left(\mathbf{P}_{B_1 B_2}\right)^{-1}$$

The expression at the end of the above equation might be easier to remember: It says that to use the matrix $\mathbf{A}_{B_2}$, you can equivalently convert coordinates from B_2 to B_1, multiply by the matrix $\mathbf{A}_{B_1}$, then convert back to coordinates in B_2. There is also the expression that "goes the other way": $\left(\mathbf{P}_{B_2 B_1}\right)^{-1} \mathbf{A}_{B_1} \left(\mathbf{P}_{B_2 B_1}\right)$. This relationship between $\mathbf{A}_{B_1}$ and $\mathbf{A}_{B_2}$ motivates the following definition:

Definition 6.3 *If A and B are square matrices then A is said to be similar to B if for some invertible matrix $\mathbf{P}$, $A = \mathbf{P}^{-1}B\mathbf{P}$.*

Note that if $A = \mathbf{P}^{-1}B\mathbf{P}$ then $B = \mathbf{P}A\mathbf{P}^{-1}$ so if we take $\mathbf{Q} = \mathbf{P}^{-1}$ then $B = \mathbf{Q}^{-1}A\mathbf{Q}$. That is, this definition is "symmetric" in the sense that if A is similar to B then B is similar to A. For this reason we will often simply say that A and B *are similar.*

Exercise 6.10 *Show that similarity is transitive. That is, if A is similar to B and B is similar to C, then A is similar to C.*

We summarize this discussion in the following theorem.

Theorem 6.9 *Suppose $T : \mathbb{R}^n \to \mathbb{R}^n$ is a linear transformation, $\mathbf{A}_{B_1}$ is the matrix representation of T with respect to one basis and $\mathbf{A}_{B_2}$ is the matrix representation of T with respect to another basis. Then $\mathbf{A}_{B_1}$ and $\mathbf{A}_{B_2}$ are similar. In fact, $\mathbf{A}_{B_1} = \mathbf{P}^{-1}\mathbf{A}_{B_2}\mathbf{P}$, where $\mathbf{P}$ is the transition matrix from B_1 to B_2. In particular, the matrix expression of a linear transformation is always similar to the matrix expression with respect to the standard basis.*

On the other hand, suppose that $\mathbf{P}$ is any $n \times n$ invertible matrix, and $B_1 = \{\mathbf{e}_1, \ldots, \mathbf{e}_n\}$ is the standard basis, and T is a linear transformation. Then according to Theorem 6.3, $B_2 := \{\mathbf{P}\mathbf{v}_1, \ldots, \mathbf{P}\mathbf{v}_n\}$ is a basis of $\mathbb{R}^n$. Since B_1 is the standard basis, $\mathbf{C}_{B_1}$ is simply the identity matrix. On the other hand, $\mathbf{C}_{B_2}$ has as its ith column the vector $\mathbf{P}\mathbf{e}_i$, which is simply the ith column of $\mathbf{P}$. That is, the transition matrix from B_2 to B_1 is just $\mathbf{C}_{B_1}^{-1}\mathbf{C}_{B_2} = \mathbf{I}^{-1}\mathbf{P} = \mathbf{P}$. In other words,

$$\mathbf{A}_{B_2} = \mathbf{P}^{-1}\mathbf{A}\mathbf{P}$$

This means that the matrix representing a T transformation of $\mathbb{R}^n$ with respect to any basis is similar to the matrix representing T with respect to the standard basis. But being similar is transitive, according to Exercise 6.10. This means that the matrices representing a transformation with respect to any two bases must be similar. We state this as a theorem:

Theorem 6.10 *Let $T : \mathbb{R}^n \to \mathbb{R}^n$ be a linear transformation. Then matrices $\mathbf{A}_1$ and $\mathbf{A}_2$ represent T with respect to some choice of bases in $\mathbb{R}^n$ if and only if $\mathbf{A}_1$ and $\mathbf{A}_2$ are similar.*

Remark 6.3 *While we will not state the theorem explicitly, essentially the same process, along with Theorem 6.9, is valid in any finite dimensional vector space V, because all such spaces are isomorphic to $\mathbb{R}^n$. That is, given a linear transformation $T : V \to V$, we can choose a basis B for V and express T as a matrix $\mathbf{A}_B$ in terms of that basis. Then another matrix $\mathbf{M}$ represents T with respect to some basis if and only if $\mathbf{A}_B$ and $\mathbf{M}$ are similar.*

In the next MP you will see the process of "diagonalization" in which some matrices can be shown to be similar to a particularly simple kind of matrix, namely a diagonal matrix. Not all matrices can be "diagonalized", and we will revisit this topic in the next chapter to find conditions that will tell you both that a matrix can be diagonalized, and how to diagonalize it when it is. For example, rotation matrices cannot be diagonalized, but at present we don't have a good way to see why that is true.

MP 67 *TRANSITION MATRIX/SIMILARITY*

6.6 Row and Column Spaces of a Matrix

As you have seen before, an $m \times n$ matrix can be viewed as a row of column vectors or a column of row vectors. That is,

$$
\mathbf{A} = \begin{bmatrix} \mathbf{A}_1 \\ \mathbf{A}_2 \\ \vdots \\ \mathbf{A}_m \end{bmatrix} \text{ or } \mathbf{A} = \begin{bmatrix} \mathbf{A}^1 & \mathbf{A}^2 & \cdots & \mathbf{A}^n \end{bmatrix}
$$

The row vectors $\mathbf{A}_i$ are vectors in $\mathbb{R}^n$, because there are n columns, and the span of $\{\mathbf{A}_1, \ldots, \mathbf{A}_m\}$ is called the *row space* of $\mathbf{A}$. Since there are m rows, the dimension of the row space is at most m. The vectors $\{\mathbf{A}^1, \ldots, \mathbf{A}^n\}$ are vectors in $\mathbb{R}^m$ because there are m rows, and their span is called the *column space* of $\mathbf{A}$. The dimension of the column space is called the *rank of* $\mathbf{A}$, which is at most n. The rank of a matrix is very important, and it can be calculated using Gauss-Jordan elimination. Consider a matrix in reduced echelon form:

$$
\mathbf{R} = \begin{bmatrix} 1 & a & 0 & b & d & 0 \\ 0 & 0 & 1 & c & e & 0 \\ 0 & 0 & 0 & 0 & 0 & 1 \\ 0 & 0 & 0 & 0 & 0 & 0 \\ 0 & 0 & 0 & 0 & 0 & 0 \end{bmatrix} \tag{6.7}
$$

The first thing to deduce, which can be checked in the particular matrix above, is that the columns that contain leading entries consist of different standard basis elements and therefore they are linearly independent. Moreover, the only possible non-zero entries in the other columns are to the right of leading entries. In fact, the columns without leading entries are all linear combinations of the columns with leading entries. For example, $\mathbf{R}^2 = a\mathbf{R}^1$ and $\mathbf{R}^5 = d\mathbf{R}^1 + e\mathbf{R}^2$. (To be clear–the superscripts refer to columns of $\mathbf{R}$ and not powers of $\mathbf{R}$, which might not even exist!) In other words, the columns with leading entries are a basis for the column space, and therefore the rank of $\mathbf{R}$ is the number of leading entries.

Now let's consider the row space of $\mathbf{R}$. Because $\mathbf{R}$ is in reduced echelon form, each leading entry is to the right of those above it, and there are only 0s in any row to the left of its leading entry (if it has one). In other words, the non-zero rows are linearly independent, meaning that the non-zero rows are a basis for the row space. Moreover, the number of non-zero rows is precisely equal to the number of leading entries.

The next question to consider is to what extent these facts are still true for a matrix $\mathbf{A}$ that is *not in reduced echelon form*. Since we can reduce $\mathbf{A}$ to reduced echelon form with row operations, the question becomes, *do row operations change the rank of a matrix, and what are the impacts on the row and column spaces?* At first the answer might not be obvious. For example, the row operation that swaps rows actually *swaps the entries in the column vectors*, which at first glance seems to be a strange thing to do to a vector! However, our theorems come to our rescue. By Theorem 2.10, every row operation on a matrix $\mathbf{A}$ is equivalently accomplished by multiplying on the left by an elementary matrix, and elementary matrices are non-singular. Let's write out the matrix multiplication, considering $\mathbf{A}$ as a row of column vectors:

$$\mathbf{EA} = \mathbf{E}\begin{bmatrix} \mathbf{A}^1 & \mathbf{A}^2 & \cdots & \mathbf{A}^n \end{bmatrix} = \begin{bmatrix} \mathbf{EA}^1 & \mathbf{EA}^2 & \cdots & \mathbf{EA}^n \end{bmatrix}$$

That is, $\mathbf{E}$ takes each column vector $\mathbf{A}^j$ to the column vector $\mathbf{EA}^j$ in $\mathbf{EA}$. Since $\mathbf{E}$ represents an isomorphism, the dimension of the span of $\{\mathbf{A}^1, \ldots \mathbf{A}^n\}$ is the same as the dimension of the span of $\{\mathbf{EA}^1, \ldots \mathbf{EA}^n\}$ by Theorem 6.3. In other words, multiplication by $\mathbf{E}$ (or any non-singular matrix) does not change the rank of $\mathbf{A}$. We collect all of our observations into an important theorem:

Theorem 6.11 *Multiplication on the left by a non-singular matrix does not change the rank of a matrix $\mathbf{A}$. In particular,*

1. *Row operations do not change the rank of a matrix $\mathbf{A}$.*

2. *The rank of $\mathbf{A}$ is the number of leading entries in its reduced echelon form.*

3. *The rank of $\mathbf{A}$ is the number of non-zero rows in its reduced echelon form.*

4. *The dimension of the row space is equal to the rank of $\mathbf{A}$.*

5. *The rank of $\mathbf{A}$ is equal to the rank of $\mathbf{A}^T$.*

Exercise 6.11 *Explain why the last statement in the above theorem is true.*

Exercise 6.12 *In the following matrix write each column without a leading entry as a linear combination of columns with leading entries.*

$$\mathbf{R} = \begin{bmatrix} 0 & 1 & -1 & 0 & 0 & 0 \\ 0 & 0 & 0 & 1 & 0 & 2 \\ 0 & 0 & 0 & 0 & 1 & 3 \\ 0 & 0 & 0 & 0 & 0 & 0 \end{bmatrix}$$

MP 68 *RANK*

We now have a very important statement about the column space:

Theorem 6.12 *If $\mathbf{A}$ is an $m \times n$ matrix then the column space of $\mathbf{A}$ is the range of the linear transformation $T_{\mathbf{A}} : \mathbb{R}^m \to \mathbb{R}^n$. In particular, the rank of $\mathbf{A}$ is the dimension of the range of $T_{\mathbf{A}}$.*

Proof. If we apply T_A to a standard basis element $\mathbf{e}_i$, as we have seen before, $T_{\mathbf{A}}(\mathbf{e}_i) = \mathbf{A}\mathbf{e}_i = \mathbf{A}^i$. Since $T_{\mathbf{A}}$ is linear, if we write $\mathbf{v} = (v_1, \ldots, v_n) = \sum_{i=1}^n v_i \mathbf{e}_i$, we see that

$$T_{\mathbf{A}}(\mathbf{v}) = \sum_{i=1}^n v_i T_{\mathbf{A}}(\mathbf{e}_i) = \sum_{i=1}^n v_i \mathbf{A}^i$$

That is, the vector $T_{\mathbf{A}}(\mathbf{v})$ is precisely a linear combination of the columns $\mathbf{A}^i$, which by definition is the column space. ∎

Exercise 6.13 *Show that swapping rows changes the column space of the matrix $\begin{bmatrix} 0 & 1 \\ 0 & 0 \end{bmatrix}$. Hint: show that $(1,0)$ is no longer in the column space after swapping rows. This shows that while row operations do not change the **dimension** of the column space (which is the rank) they can change the column space itself.*

The above exercise should be contrasted with the next theorem:

Theorem 6.13 *If $\mathbf{A}$ is a matrix then row operations do not change the row space of $\mathbf{A}$. In particular the row space of $\mathbf{A}$ is the same as the row space of the reduced echelon form of $\mathbf{A}$.*

Proof. Swapping rows simply changes the order in which the vectors are listed, but doesn't change their span. Suppose that we replace the row $\mathbf{A}_j$ by $\mathbf{A}'_j = \mathbf{A}_j + c\mathbf{A}_k$ for some scalar c and row $\mathbf{A}_k$. If $\mathbf{w}$ is in the row space then $\mathbf{w} = \sum_{i=1}^m w_i \mathbf{A}_i$. But we can write

$$\mathbf{w} = \sum_{i=1}^m w_i \mathbf{A}_i = \sum_{i \neq j}^m w_i \mathbf{A}_i + w_j(\mathbf{A}'_j - c\mathbf{A}_k)$$

$$= \sum_{i \neq j, i \neq k} w_i \mathbf{A}_i + w_j \mathbf{A}'_j + (w_k - c)\mathbf{A}_k$$

That is, $\mathbf{w}$ is still in the span of the rows of the new matrix. This shows that the row space of $\mathbf{A}$ is contained in the row space of the new matrix. But this row operation can be reversed with another row operation, which implies that the row space of the new matrix is contained in the row space of $\mathbf{A}$—meaning that, the row space is unchanged by the second type of row operation. The rest of the proof is an exercise. ∎

Exercise 6.14 *Finish the proof of Theorem 6.13 by showing that if a row of* **A** *is multiplied by a non-zero scalar then the row space of* **A** *is contained in the row space of the new matrix. Since this row operation can be reversed by the same type of row operation, this shows that multiplication of a row by a non-zero scalar doesn't change the row space. Hint: use the above proof for the second row operation as a model–but the proof for this exercise involves a simpler calculation.*

Note that Theorem 6.13 has another application, namely finding the dimension of the subspace W spanned by a set $S = \{\mathbf{v}_1, \ldots, \mathbf{v}_r\}$. If S is linearly independent, then of course we know that $\dim W = r$, but otherwise we only know that $\dim W < r$. To find the actual dimension, we create the matrix $\mathbf{C}_S$, reduce it, and count the number of leading entries.

MP 69 *DIMENSION OF SUBSPACE*

Of course you could also put the vectors in S into the matrix as rows, which would have the advantage of preserving the span according to Theorem 6.13, but I am emphasizing the matrix $\mathbf{C}_S$ for consistency and to make the processes easier to remember.

6.7   Null Space of a Matrix

The *null space* of a matrix is the solution set of the homogeneous system $\mathbf{AX} = \mathbf{0}$, which you should recognize is also the kernel of the linear transformation $T_\mathbf{A}$. Let's consider the following matrix in reduced echelon form:

$$\mathbf{R} = \begin{bmatrix} 1 & a & 0 & b & 0 & i \\ 0 & 0 & 1 & c & 0 & j \\ 0 & 0 & 0 & 0 & 1 & k \\ 0 & 0 & 0 & 0 & 0 & 0 \\ 0 & 0 & 0 & 0 & 0 & 0 \end{bmatrix}$$

The corresponding homogeneous system of equations $\mathbf{RX} = \mathbf{0}$ is

$$x_1 + ax_2 + bx_4 + ix_6 = 0$$
$$x_3 + cx_4 + jx_6 = 0$$
$$x_5 + kx_6 = 0$$

As you have learned, we may parameterize the solutions using the free variables, which correspond to columns with no leading entries: $x_2 = r, x_4 = s$, $x_6 = t$. We may then solve for the leading variables and obtain parameterized solutions:

$$
\begin{bmatrix} x_1 \\ x_2 \\ x_3 \\ x_4 \\ x_5 \end{bmatrix} = \begin{bmatrix} -(r+bs+it) \\ r \\ -(cs+jt) \\ -kt \\ t \end{bmatrix} = r \begin{bmatrix} -1 \\ 1 \\ 0 \\ 0 \\ 0 \end{bmatrix} + s \begin{bmatrix} -b \\ 0 \\ -c \\ 0 \\ 0 \end{bmatrix} + t \begin{bmatrix} -i \\ 0 \\ -j \\ -k \\ -1 \end{bmatrix} \tag{6.8}
$$

We may now interpret the parameterized solutions in a new way: they describe solutions to the homogeneous system, i.e. the null space, as a linear combination of the three vectors on the right:

$$
\begin{bmatrix} -1 \\ 1 \\ 0 \\ 0 \\ 0 \end{bmatrix} \begin{bmatrix} -b \\ 0 \\ -c \\ 0 \\ 0 \end{bmatrix} \begin{bmatrix} -i \\ 0 \\ -j \\ -k \\ -1 \end{bmatrix}
$$

That is, those three vectors *span* the null space. Are they a basis? In this particular example, these column vectors are linearly independent. For example, the second entry of the first vector is 1, but the second entries of the other two vectors are 0, the first cannot be a linear combination of the other two. Similarly, the bottom entries shows that the third cannot be a linear combination of the other two.

Exercise 6.15 *Explain why the middle of the above three vectors is not a linear combination of the other two. This is a little more complicated than the other two arguments, but it again comes down to leveraging the entries that are 0. You must explain this without resorting to using a computer!*

From the construction of the parameterized system there is no obvious way to see that the resulting vectors are always linearly independent. Instead, we will use the following extremely important theorem about dimension.

Theorem 6.14 *(Dimension Sum Theorem) If $T : V \to W$ is a linear transformation between finite dimensional vector spaces, then the dimension of V is equal to the sum of the dimension of $\ker T$ plus the dimension of the range of T.*

Proof. Let $\{v_1, \ldots, v_k\}$ be a basis for $\ker T$. By the Basis Extension Theorem we can extend this basis to a basis $B = \{v_1, \ldots, v_k, w_1, \ldots, w_r\}$ of V. Since the dimension of V is $k + r$, we will be finished if we show that $\{T(w_1), \ldots, T(w_r)\}$ is a basis for the range of T. Vectors in the range of T are of the form

$$
T(c_1 v_1 + \cdots + c_k v_k + d_1 w_1 + \cdots + d_r w_r)
$$

$$
= c_1 T(v_1) + \cdots + c_k T(v_k) + d_1 T(w_1) + \cdots + d_r T(w_r)
$$

$$
= 0 + \cdots + 0 + d_1 T(w_1) + \cdots + d_r T(w_r) = d_1 T(w_1) + \cdots + d_r T(w_r)
$$

This shows that the vectors $\{T(\mathbf{w}_1), \ldots, T(\mathbf{w}_r)\}$ span the range of T. Now suppose we have a linear combination

$$\sum_{i=1}^{r} k_i T(\mathbf{w}_i) = \mathbf{0}$$

The proof will be finished by Theorem 5.8 if we can show that $k_i = 0$ for all i. But consider the vector

$$\mathbf{w} = \sum_{i=1}^{r} k_i \mathbf{w}_i = k_1 \mathbf{w}_1 + \cdots + k_r \mathbf{w}_r + 0\mathbf{v}_1 + \cdots + 0\mathbf{v}_k \qquad (6.9)$$

in V. By linearity of T,

$$T(\mathbf{w}) = T\left(\sum_{i=1}^{r} k_i \mathbf{w}_i\right) = \sum_{i=1}^{r} k_i T(\mathbf{w}_i) = \mathbf{0}$$

This means that $\mathbf{w}$ is in $\ker T$, and so we can also write

$$\mathbf{w} = \sum_{j=1}^{k} c_j \mathbf{v}_j = 0\mathbf{w}_1 + \cdots + 0\mathbf{w}_r + c_1 \mathbf{v}_1 + \cdots + c_k \mathbf{v}_k \qquad (6.10)$$

Equations (6.9) and (6.10) both express $\mathbf{w}$ as a linear combination of the vectors in the basis B, so by uniqueness they must have the same coefficients. But this means that $k_i = 0$ for all i, finishing the proof that the vectors $\{T(\mathbf{w}_1), \ldots, T(\mathbf{w}_r)\}$ are linearly independent. ∎

Theorem 6.15 *The process to parameterize solutions of a homogeneous system of equations via Gauss-Jordan elimination expresses those solutions as a linear combination of a basis of the null space.*

Proof. For this proof we will use the discussion about the null space, along with Theorem 6.11. If a matrix in reduced echelon form has n columns, then the r columns with leading entries form a basis for the column space, and each of the remaining $n - r$ columns corresponds to a free variable. As discussed concerning (6.8), the null space is spanned by a set of vectors, one each for each free column. Since the dimension of the null space is $n - r$ by Theorem 6.14, those vectors must be linearly independent, hence a basis for the null space. ∎

MP 70 *BASIS FOR NULL SPACE*

6.8 Orthonormal Bases and Orthogonal Matrices

As I have mentioned previously, it is often possible to find bases that greatly simplify calculations and can also help prove theorems. In this section we will

consider an important type of basis, namely an *orthonormal basis*s. This type
of basis not only makes some kinds of calculations easier, it has important
mathematical consequences that can be used to understand the relationship
between the null, column and row spaces of a matrix.

Definition 6.4 *A set of non-zero vectors $S = \{\mathbf{v}_1, \ldots, \mathbf{v}_k\}$ in $\mathbb{R}^n$ is said to be
orthogonal if when $i \neq j$, $\mathbf{v}_i$ and $\mathbf{v}_j$ are orthogonal. S is said to be orthonormal
if in addition, each $\mathbf{v}_i$ is a unit vector.*

The word "orthonormal" is derived from "orthogonal" and "normal",
which refers to unit vectors. Put another way, if S is orthonormal, $\mathbf{v}_i \cdot \mathbf{v}_j = 0$
when $i \neq j$ and $\mathbf{v}_i \cdot \mathbf{v}_i = 1$ for all i. For example, the standard basis $\{\mathbf{e}_1, \ldots, \mathbf{e}_n\}$
is orthonormal. Notation can be greatly simplified using something called the
Kronecker Delta, defined by $\delta_{ij} = 0$ when $i \neq j$ and $\delta_{ii} = 1$ for all i. For
example, the identity matrix is precisely $[\delta_{ij}]$, and orthonormality of a set
$S = \{\mathbf{v}_1, \ldots, \mathbf{v}_k\}$ of vectors means precisely that $\mathbf{v}_i \cdot \mathbf{v}_j = \delta_{ij}$.

Theorem 6.16 *Orthogonal vectors are linearly independent.*

Proof. Suppose $\{\mathbf{v}_1, \ldots, \mathbf{v}_n\}$ are orthogonal and suppose

$$\sum_{i=1}^n k_i \mathbf{v}_i = \mathbf{0}$$

We need to show that $k_j = 0$ for all j. By bilinearity of the dot product we
know that for any j,

$$0 = \left(\sum_{i=1}^n k_i \mathbf{v}_i\right) \cdot \mathbf{v}_j = \sum_{i=1}^n k_i (\mathbf{v}_i \cdot \mathbf{v}_j) = k_j (\mathbf{v}_j \cdot \mathbf{v}_j) = k_j$$

Since $\mathbf{v}_j \neq \mathbf{0}$, by positive definiteness $(\mathbf{v}_j \cdot \mathbf{v}_j) \neq 0$, which implies $k_j = 0$. ∎
The next theorem shows that when you have an orthonormal basis, the
dot product of any two vectors is computed in exactly the same way as with
the standard basis, using the coordinates of the vectors with respect to the
basis. This can greatly simplify calculations.

Theorem 6.17 *Suppose that $S = \{\mathbf{v}_1, \ldots, \mathbf{v}_n\}$ is an orthonormal basis for
$\mathbb{R}^n$. For any $\mathbf{v} = \sum_{i=1}^n k_i \mathbf{v}_i$ and $\mathbf{w} = \sum_{j=1}^n c_j \mathbf{v}_j$,*

$$\mathbf{v} \cdot \mathbf{w} = k_1 c_1 + \cdots + k_n c_n = \sum_{i=1}^n k_i c_i$$

Proof. I used different indexing variables for $\mathbf{v}$ and $\mathbf{w}$ because we are about to compute the dot product using bilinearity.

$$\mathbf{v} \cdot \mathbf{w} = \left(\sum_{i=1}^{n} k_i \mathbf{v}_i \right) \cdot \left(\sum_{j=1}^{n} c_j \mathbf{v}_j \right) = \sum_{i=1}^{n} \sum_{j=1}^{n} k_i c_j \left(\mathbf{v}_i \cdot \mathbf{v}_j \right)$$

$$= \sum_{i=1}^{n} \sum_{j=1}^{n} k_i c_j \delta_{ij} = \sum_{i=1}^{n} k_i c_i$$

■

Exercise 6.16 *Show that $\sum_{i=1}^{n} \sum_{j=1}^{n} k_i c_j \delta_{ij} = \sum_{i=1}^{n} k_i c_i$ to justify the last equality in the above calculation. This is not complicated–just use the definition of the Kronecker Delta δ_{ij}.*

Other applications of orthonormality involve a very important type of matrix.

Definition 6.5 *A invertible matrix $\mathbf{A}$ is called orthogonal if $\mathbf{A}^T = \mathbf{A}^{-1}$.*

Checking whether a (square) matrix is orthogonal is relatively easy: just multiply it by its transpose to see if you get the identity.

Example 6.1 *For rotations we have*

$$\mathbf{R}_\theta \mathbf{R}_\theta^T = \begin{bmatrix} \cos\theta & \sin\theta \\ -\sin\theta & \cos\theta \end{bmatrix} \begin{bmatrix} \cos\theta & -\sin\theta \\ \sin\theta & \cos\theta \end{bmatrix}$$

$$= \begin{bmatrix} \cos^2\theta + \sin^2\theta & -\sin\theta\cos\theta + \sin\theta\cos\theta \\ -\cos\theta\sin\theta + \cos\theta\sin\theta & \cos^2\theta + \sin^2\theta \end{bmatrix} = \begin{bmatrix} 1 & 0 \\ 0 & 1 \end{bmatrix}$$

That is, rotations are orthogonal.

Exercise 6.17 *Check that the matrices $\mathbf{M}_1$ and $\mathbf{M}_2$ in Exercise 1.1 are orthogonal.*

Exercise 6.18 *Let $\mathbf{A}$ and $\mathbf{B}$ be orthogonal.*

1. *Show that $\mathbf{A}^T$ and $\mathbf{A}^{-1}$ are orthogonal.*
2. *Show that $\mathbf{AB}$ is orthogonal.*

The next theorem explains the word "orthogonal" in this context.

Theorem 6.18 *An $n \times n$ matrix $\mathbf{A}$ is orthogonal if and only if the columns of $\mathbf{A}$ are an orthonormal basis for $\mathbb{R}^n$. In particular, a set $S = \{\mathbf{v}_1, \ldots, \mathbf{v}_n\}$ of vectors in $\mathbb{R}^n$ is an orthonormal basis for $\mathbb{R}^n$ if and only if $\mathbf{C}_S$ is an orthogonal matrix.*

Proof. We will regard $\mathbf{A}$ as a as a row of columns:

$$\mathbf{A} = \begin{bmatrix} \mathbf{A}^1 & \cdots & \mathbf{A}^n \end{bmatrix}$$

We will regard $\mathbf{A}^T$ as a column of rows–but of course its rows are the columns of $\mathbf{A}$:

$$\mathbf{A}^T = \begin{bmatrix} \mathbf{A}^1 \\ \vdots \\ \mathbf{A}^n \end{bmatrix}$$

Now

$$\mathbf{I} = \begin{bmatrix} 1 & \cdots & 0 & \cdots & 1 \\ \vdots & \ddots & \vdots & \vdots & \vdots \\ 0 & \cdots & 1 & \cdots & 0 \\ \vdots & \vdots & \vdots & \ddots & \vdots \\ 0 & \cdots & 0 & \cdots & 1 \end{bmatrix}$$

and

$$\left(\mathbf{A}^T\right)\mathbf{A} = \begin{bmatrix} \mathbf{A}^1 \cdot \mathbf{A}^1 & \cdots & \mathbf{A}^1 \cdot \mathbf{A}^i & \cdots & \mathbf{A}^1 \cdot \mathbf{A}^n \\ \vdots & \ddots & \vdots & \vdots & \vdots \\ \mathbf{A}^i \cdot \mathbf{A}^1 & \cdots & \mathbf{A}^i \cdot \mathbf{A}^i & \cdots & \mathbf{A}^i \cdot \mathbf{A}^n \\ \vdots & \vdots & \vdots & \ddots & \vdots \\ \mathbf{A}^n \cdot \mathbf{A}^1 & \cdots & \mathbf{A}^n \cdot \mathbf{A}^i & \cdots & \mathbf{A}^n \cdot \mathbf{A}^n \end{bmatrix}$$

Comparing entries of these matrices shows that $\mathbf{A}$ is orthogonal if and only if its columns are orthonormal. ■

Exercise 6.19 *Prove the version of Theorem 6.18 that replaces "columns" by "rows". Hint: Do not rework the above proof (unless you just want to practice)! Just apply the case that was proved to $\mathbf{A}^T$.*

Suppose that $\mathbf{A}$ is orthogonal. Then

$$\det(\mathbf{A})\det(\mathbf{A}^T) = \det(\mathbf{A}\mathbf{A}^T) = \det \mathbf{I} = 1$$

But $\det(\mathbf{A}) = \det\left(\mathbf{A}^T\right)$ and therefore $(\det \mathbf{A})^2 = 1$. This proves the following theorem:

Theorem 6.19 *If $\mathbf{A}$ is an orthogonal matrix then $\det \mathbf{A} = \pm 1$.*

Theorem 6.20 *The transition matrix from one orthonormal basis B_1 of $\mathbb{R}^n$ to another orthonormal basis B_2 of $\mathbb{R}^n$ is orthogonal.*

Proof. Note that the transition function from B_1 to the standard basis is just $\mathbf{C}_{B_1}$, which has as its columns the vectors in B_1. By Theorem 6.18 $\mathbf{C}_{B_1}$ is orthogonal. The same is true for $\mathbf{C}_{B_2}$, and since $\mathbf{P}_{B_1 B_2} = \mathbf{C}_{B_2}^{-1} \mathbf{C}_{B_1}$, the proof is finished by Exercise 6.18. ■

The next theorem has important applications to physics, among other fields. It says that multiplication by an orthogonal matrix is a "rigid motion", also called an "isometry" of Euclidean space. The theorem says that multiplication by orthogonal matrices "preserves the dot product". Since the dot product is preserved, so are lengths $\|\mathbf{v}\| = \sqrt{\mathbf{v} \cdot \mathbf{v}}$ of vectors, and therefore distances between points. The word "isometry" comes from the prefix "iso", which as you have seen before generally means "the same" and "metric", which means the measure of something—in this case the distance between points.

Theorem 6.21 *If $\mathbf{A}$ is orthogonal and $\mathbf{v}, \mathbf{w}$ are vectors, then $\mathbf{Av} \cdot \mathbf{Aw} = \mathbf{v} \cdot \mathbf{w}$.*

Exercise 6.20 *Prove Theorem 6.21 using Formula (4.17)—this is a 1-line proof also using the definition of "orthogonal".*

6.9 Orthogonal Subspaces and Gram-Schmidt

Definition 6.6 *If V and W are subspaces of a vector space, the span of V and W consists of the set of all $\mathbf{v} + \mathbf{w}$ such that $\mathbf{v}$ is in V and $\mathbf{w}$ is in W. We say that V and W are orthogonal if whenever $\mathbf{v}$ is in V and $\mathbf{w}$ is in W, $\mathbf{v} \cdot \mathbf{w} = 0$. We write $V \perp W$.*

Theorem 6.22 *Suppose V and W are subspaces of $\mathbb{R}^n$ and $V \perp W$. Then*

1. $V \cap W = \{\mathbf{0}\}$

2. If $\{\mathbf{v}_1, \ldots, \mathbf{v}_m\}$ is a linearly independent subset of V and $\{\mathbf{w}_1, \ldots, \mathbf{w}_k\}$ is a linearly independent subset of W, then $\{\mathbf{v}_1, \ldots, \mathbf{v}_m, \mathbf{w}_1, \ldots, \mathbf{w}_k\}$ is linearly independent.

3. If $\mathbf{u}$ is in the span of V and W then there is a unique way to write $\mathbf{u} = \mathbf{v} + \mathbf{w}$, where $\mathbf{v}$ is in V and $\mathbf{w}$ is in W.

Proof. If $\mathbf{u}$ is in $V \cap W$, then by definition $\mathbf{u}$ is in V and $\mathbf{u}$ is in W. Since V and W are orthogonal, this means that

$$0 = \mathbf{u} \cdot \mathbf{u} = \|\mathbf{u}\|^2$$

By positive definiteness, $\mathbf{u} = \mathbf{0}$.

For the second part, suppose

$$\sum_{i=1}^{m} k_i \mathbf{v}_i + \sum_{j=1}^{k} c_i \mathbf{w}_j = \mathbf{0}$$

We need to prove that $k_i = 0$ for all i and $c_j = 0$ for all j. Since the vector on the right is $\mathbf{0}$, its dot product with itself is 0. Referring back to the calculation (4.14) involving bilinearity, we have

$$0 = \left\| \sum_{i=1}^{m} k_i \mathbf{v}_i \right\|^2 + \left\| \sum_{j=1}^{k} c_j \mathbf{w}_j \right\|^2 + 2 \left(\sum_{i=1}^{m} k_i \mathbf{v}_i \right) \cdot \left(\sum_{j=1}^{k} c_j \mathbf{w}_j \right)$$

Now using bilinearity on the last term, together with the fact that $V \perp W$, we see that

$$\left(\sum_{i=1}^{m} k_i \mathbf{v}_i \right) \cdot \left(\sum_{j=1}^{k} c_j \mathbf{w}_j \right) = \sum_{i,j} k_i c_j (\mathbf{v}_i \cdot \mathbf{w}_j) = 0$$

This means that

$$\left\| \sum_{i=1}^{m} k_i \mathbf{v}_i \right\|^2 + \left\| \sum_{j=1}^{k} c_j \mathbf{w}_j \right\|^2 = 0$$

Positive definiteness implies that $\sum_{i=1}^{m} k_i \mathbf{v}_i = \mathbf{0}$, and since $\{\mathbf{v}_1, \ldots, \mathbf{v}_m\}$ was linearly independent, $k_i = 0$ for all i. Similarly, $c_i = 0$ for all i.

For the third part, suppose that we can write $\mathbf{u} = \mathbf{v}_1 + \mathbf{w}_1 = \mathbf{v}_2 + \mathbf{w}_2$ with $\mathbf{v}_1$ and $\mathbf{v}_2$ in V, and $\mathbf{w}_1$ and $\mathbf{w}_2$ in W. Then

$$(\mathbf{v}_1 - \mathbf{v}_2) + (\mathbf{w}_1 - \mathbf{w}_2) = \mathbf{u} - \mathbf{u} = \mathbf{0}$$

From this point, showing that $\mathbf{v}_1 = \mathbf{v}_2$ and $\mathbf{w}_1 = \mathbf{w}_2$, as required, is an exercise. ∎

Remark 6.4 *You should check that the definition of orthogonal subspace and the proof of the above theorem only requires the three basic properties of the dot product: positive definiteness, symmetry, and bilinearity. Therefore the proof is valid in more general vector spaces that are equipped with such product, which is known as a positive definite, symmetric bilinear form , or an inner product.*

Exercise 6.21 *Finish the proof of Theorem 6.22 by expanding the dot product of $(\mathbf{v}_1 - \mathbf{v}_2) + (\mathbf{w}_1 - \mathbf{w}_2)$ with itself (which must be 0), using the fact that $V \perp W$, and showing that $\mathbf{v}_1 = \mathbf{v}_2$ and $\mathbf{w}_1 = \mathbf{w}_2$.*

At this point it should seem reasonable to you that if V is a subspace of $\mathbb{R}^n$, it would be useful to find some subspace W such that $V \perp W$ and V, W span $\mathbb{R}^n$.

Definition 6.7 *For any nonempty subset S of $\mathbb{R}^n$, the orthogonal complement of S, called $S^\perp$ is the set of all vectors $\mathbf{w}$ in $\mathbb{R}^n$ such that for any vector $\mathbf{v}$ in S, $\mathbf{w} \perp \mathbf{v}$.*

Exercise 6.22 *Show that for any nonempty set S (whether or not it is a subspace!) $S^\perp$ is a subspace of $\mathbb{R}^n$.*

Theorem 6.23 *For any subspace V of $\mathbb{R}^n$, V and $V^\perp$ span $\mathbb{R}^n$. Conversely, if V and W span $\mathbb{R}^n$ and $V \perp W$ then $W = V^\perp$.*

This important theorem will be proved using a process known as the Gram-Schmidt orthogonalization process to find an orthonormal basis for any subspace V of $\mathbb{R}^n$. Let $\{\mathbf{v}_1, \ldots, \mathbf{v}_m\}$ be a basis for V. Start with the unit vector $\mathbf{u}_1 = \frac{\mathbf{v}_1}{\|\mathbf{v}_1\|}$, so $\mathbf{u}_1$ has the same span as $\mathbf{v}_1$. Let $\mathbf{u}_2$ be the unit vector in the direction of $\mathbf{w}_2 := \mathbf{v}_2 - proj_{\mathbf{v}_1}(\mathbf{v}_2)$.

Exercise 6.23 *Show that $\mathbf{w}_2 \perp \mathbf{u}_1$, $\mathbf{w}_2 \neq \mathbf{0}$, and the span of $\mathbf{u}_1, \mathbf{u}_2$ is equal to the span of $\mathbf{v}_1, \mathbf{v}_2$.*

If V is two-dimensional then we are done. If not, there is some $\mathbf{w}$ in V that is not in the span of $\mathbf{u}_1$ and $\mathbf{v}_2$. The previous case was a bit more straightforward since you were able to simply use the projection of $\mathbf{v}_2$ onto the span of $\mathbf{v}_1$. In this case, we have no definition of the projection of $\mathbf{w}$ onto the span of $\mathbf{u}_1$ and $\mathbf{u}_2$ (in fact we will use Gram-Schmidt to *define* this projection later). At any rate, we need to find a vector $\mathbf{w}_3 = a_1\mathbf{u}_1 + a_2\mathbf{u}_2 + a\mathbf{w}$ such that $\mathbf{w}_3 \perp \mathbf{u}_2$, $\mathbf{w}_3 \perp \mathbf{u}_1$, and $a \neq 0$. Once we have done so, we may let $\mathbf{u}_3 := \frac{\mathbf{w}_3}{\|\mathbf{w}_3\|}$ and continue the process. We have two equations, using the fact that $\{\mathbf{u}_1, \mathbf{u}_2\}$ is orthonormal:

$$0 = \mathbf{w}_3 \cdot \mathbf{u}_1 = a_1\mathbf{u}_1 \cdot \mathbf{u}_1 + a_2\mathbf{u}_2 \cdot \mathbf{u}_1 + a\mathbf{w} \cdot \mathbf{u}_1 = a_1 + a\mathbf{w} \cdot \mathbf{u}_1$$

$$0 = \mathbf{w}_3 \cdot \mathbf{u}_2 = a_1\mathbf{u}_1 \cdot \mathbf{u}_2 + a_2\mathbf{u}_2 \cdot \mathbf{u}_2 + a\mathbf{w} \cdot \mathbf{u}_2 = a_2 + a\mathbf{w} \cdot \mathbf{u}_2$$

This is a homogeneous linear system with variables a_1, a_2, a:

$$\begin{bmatrix} a_1 & 0 & (\mathbf{w} \cdot \mathbf{u}_1)\,a & | & 0 \\ 0 & a_2 & (\mathbf{w} \cdot \mathbf{u}_2)\,a & | & 0 \end{bmatrix}$$

This system has non-zero solutions with free variable a. Choosing $a = 1$ gives us

$$\mathbf{w}_3 = \mathbf{w} - (\mathbf{w} \cdot \mathbf{u}_1)\,\mathbf{u}_1 - (\mathbf{w} \cdot \mathbf{u}_2)\,\mathbf{u}_2$$

and we take $\mathbf{u}_3 := \frac{\mathbf{w}_3}{\|\mathbf{w}_3\|}$.

Exercise 6.24 *Write out the matrix that allows you to find $\mathbf{w}_4$ in the Gram-Schmidt process and write down the formula for $\mathbf{w}_4$.*

Exercise 6.25 *Write out the general form of $\mathbf{w}_n$ in the Gram-Schmidt process in terms of*

  1. the vectors $\mathbf{u}_1, \ldots, \mathbf{u}_n$ and

  2. the vectors $\mathbf{v}_1, \ldots \mathbf{v}_n$, which is the way in which the Gram-Schmidt process is often described.

You do not need to write out a formal proof of your equations.

Returning to the proof of Theorem 6.23, use Gram-Schmidt to find an orthonormal basis $\mathbf{v}_1, \ldots \mathbf{v}_k$ for V, and then continue the process to extend to an orthonormal basis $B = \{\mathbf{v}_1, \ldots, \mathbf{v}_n\}$ of $\mathbb{R}^n$. For any $\mathbf{w}$ in $\mathbb{R}^n$ we may write $\mathbf{w} = \sum_{i=1}^{k} w_i \mathbf{v}_i + \sum_{i=k+1}^{n} w_i \mathbf{v}_i$. Since B is orthonormal, all of the $\mathbf{v}_i$ in the second sum are in $V^\perp$, showing that V and $V^\perp$ span $\mathbb{R}^n$.

Conversely, suppose that $W \perp V$ and V, W span $\mathbb{R}^n$. Then by definition of $V^\perp$ it is certainly true that $W \subset V^\perp$. We now use Gram-Schmidt to extend B to an orthonormal basis $\{\mathbf{v}_1, \ldots, \mathbf{v}_k, \mathbf{w}_{k+1}, \ldots, \mathbf{w}_m\}$ of the span of V and W. But by assumption, this span is $\mathbb{R}^n$, and therefore $m = n$, and $m = n - k$. That is, W has the same dimension as $V^\perp$ and is contained in $V^\perp$. Therefore $W = V^\perp$ (see Theorem 6.5).

Theorem 6.24 *If V is a subspace of $\mathbb{R}^n$ then $\left(V^\perp\right)^\perp = V$.*

Proof. Apply Theorem 6.23 with $V^\perp$ playing the role of V in the hypothesis and with V playing the role of W. From Theorem 6.23 we know that $V^\perp$ and $W = V$ span $\mathbb{R}^n$, and the conclusion of the theorem is that $(V^\perp)^\perp = W = V$. ∎

Definition 6.8 *Suppose that V is a subspace of $\mathbb{R}^n$. By Theorem 6.22 we may uniquely write $\mathbf{u} = \mathbf{v} + \mathbf{w}$, where $\mathbf{v}$ is in V and $\mathbf{w}$ is in $V^\perp$. The vector $\mathbf{v}$ is called the orthogonal projection of $\mathbf{u}$ onto V denoted by $proj_V(\mathbf{u})$.*

Exercise 6.26 *Show that if $B = \{\mathbf{u}_1, \ldots, \mathbf{u}_k\}$ is an orthonormal subset of $\mathbb{R}^n$ with span V, and $\mathbf{w}$ is a vector, then*

$$proj_V(\mathbf{w}) = (\mathbf{w} \cdot \mathbf{u}_1)\,\mathbf{u}_1 + \cdots + (\mathbf{w} \cdot \mathbf{u}_k)\,\mathbf{u}_k = \sum_{i=1}^{k} proj_{\mathbf{u}_i}(\mathbf{w})$$

Hint: Use Theorem 6.22.3. That is, you can write $\mathbf{w}$ uniquely as the sum of a vector in V and a vector in $V^\perp$. Now you can write

$$\mathbf{w} = proj_V(\mathbf{w}) + (\mathbf{w} - proj_V(\mathbf{w}))$$

and

$$\mathbf{w} = \sum_{i=1}^{k} proj_{\mathbf{u}_i}(\mathbf{w}) + (\mathbf{w} - \sum_{i=1}^{k} proj_{\mathbf{u}_i}(\mathbf{w}))$$

So you need only check that in each expression, one summand is in V and the other is in $V^\perp$.

Gram-Schmidt now has an even more compact formulation as an iterative process. As the base case, start with any unit vector $\mathbf{u}_1$. For the iterative step, suppose we have found an orthonormal set $S_k = \{\mathbf{u}_1, \ldots, \mathbf{u}_k\}$, if S_k spans the subspace in question, we are done. If not, let $\mathbf{w}$ be any vector not in the span of S_k and define

$$\mathbf{u}_{k+1} := \mathbf{w} - \sum_{i=1}^{k} proj_{\mathbf{u}_i}(\mathbf{w})$$

Exercise 6.27 *Justify the above description of Gram-Schmidt by showing that $\{\mathbf{u}_1, \ldots, \mathbf{u}_k, \mathbf{u}_{k+1}\}$ is orthonormal.*

6.10   The Structure of the Row, Column and Null Spaces

Theorem 6.25 *If $\mathbf{A}$ is an $m \times n$ matrix then the null space N and the row space R of $\mathbf{A}$ are orthogonal subspaces of $\mathbb{R}^n$. Moreover,*

> *1. $N \cap R = \{\mathbf{0}\}$*
>
> *2. If $\{\mathbf{n}_1, \ldots, \mathbf{n}_a\}$ is a basis for N and $\{\mathbf{r}_1, \ldots, \mathbf{r}_k\}$ is a basis for R (so k is the rank of $\mathbf{A}$ and a is the nullity of $\mathbf{A}$) then $\{\mathbf{n}_1, \ldots, \mathbf{n}_a, \mathbf{r}_1, \ldots, \mathbf{r}_k\}$ is a basis for $\mathbb{R}^n$.*
>
> *3. Every $\mathbf{u}$ in $\mathbb{R}^n$ may be written uniquely as a sum $proj_N(\mathbf{u}) + proj_R(\mathbf{u})$.*
>
> *4. $N = R^\perp$ and $R = N^\perp$*

Proof. The fact that $N \perp R$ is simply a restatement of Theorem 4.8. From Theorem 6.22 we obtain both that $N \cap R = \{\mathbf{0}\}$ and that $\{\mathbf{n}_1, \ldots, \mathbf{n}_a, \mathbf{r}_1, \ldots, \mathbf{r}_k\}$ is linearly independent. But from the Dimension Sum Theorem (Theorem 6.14) we know that $k + a = n$, and since $\{\mathbf{n}_1, \ldots, \mathbf{n}_a, \mathbf{r}_1, \ldots, \mathbf{r}_k\}$ is linearly independent, it must be a basis. The last statement is a consequence of Theorem 6.23. ∎

We now have a very nice picture of the three main spaces for any linear transformation $T_\mathbf{A}$ from $\mathbb{R}^n$ to $\mathbb{R}^m$. We have that $N = \ker T_\mathbf{A}$ and the row space R of $\mathbf{A}$ are orthogonal and span the domain of $T_\mathbf{A}$. $T_\mathbf{A}(R)$ is the column space C of $\mathbf{A}$. Since $R \cap N = \{\mathbf{0}\}$, $\mathbf{0}$ is the only vector in R that $T_\mathbf{A}$ takes to $\mathbf{0}$. Put another way, if consider the linear map L that is the restriction of $T_\mathbf{A}$ to the subspace R, then L is a linear isomorphism from the row space R to the column space C. Every vector $\mathbf{u}$ in $\mathbb{R}^n$ can be uniquely written as $\mathbf{u} = \mathbf{n} + \mathbf{r}$ such that $T_A(\mathbf{n}) = \mathbf{0}$ and $\mathbf{r}$ is in the row space.

Now consider the matrix $\mathbf{B} = \mathbf{A}^T$. Then $T_\mathbf{B} : \mathbb{R}^m \to \mathbb{R}^n$. The row and column spaces of $\mathbf{B}$ are the column and row spaces of $\mathbf{A}$, respectively, and we may apply the same analysis to $\mathbf{B}$: the restriction of $T_\mathbf{B}$ to C takes C isomorphically to R. What about the $\ker T_\mathbf{B}$, which is the null space N' of $\mathbf{B}$, and is a subspace of $\mathbb{R}^m$? This is yet a fourth subspace of interest. It is orthogonal to the column space of $\mathbf{A}$, and it represents those vectors that are sent to $\mathbf{0}$ by $T_\mathbf{B}$. Together these four spaces span both the domain and

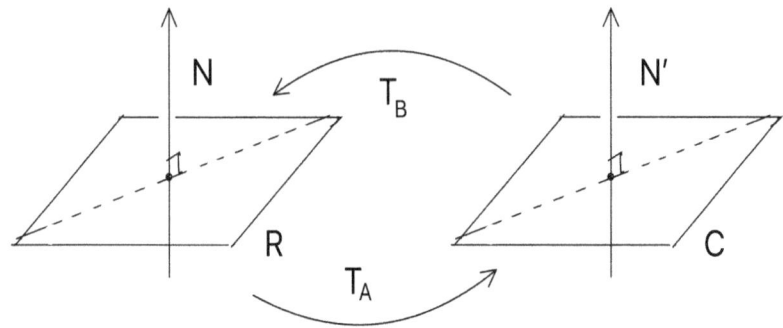

FIGURE 6.1
Row and Column Space

codomain of the transformation. Figure 6.1 illustrates the situation. However, there are some caveats concerning this picture. The limits of a two-dimensional drawing prevent showing high dimensions in any visually reasonable way. In this figure, the isomorphic subspaces R and C are shown as two-dimensional, but of course their dimension could be higher or lower than 2 (including 0). But their dimensions are always equal to each other, and equal to the rank r of $\mathbf{A}$, which is equal to the rank of $\mathbf{B} = \mathbf{A}^T$. Both N and N' are shown as one-dimensional to make the picture less cluttered, but it is important to remember that N and N' *need not have the same dimension*, despite the picture. In fact $\dim N = n - k$ and $\dim N' = m - k$, so unless $n = m$ (i.e. $\mathbf{A}$ is a square matrix), the dimensions of N and N' are not equal.

Finally, it is possible that either $N = \{\mathbf{0}\}$ or $N' = \{\mathbf{0}\}$. By Theorem 6.1, $N = \{\mathbf{0}\}$ means that $T_{\mathbf{A}}$ is $1-1$ on $\mathbb{R}^n$ and therefore is a linear isomorphism from R onto C. If both $N = \{\mathbf{0}\}$ and $N' = \{\mathbf{0}\}$ then $m = n$ and $T_A : \mathbb{R}^n \to \mathbb{R}^n$ is a linear isomorphism, as is $T_{\mathbf{B}}$. Note that this does not necessarily imply that $\mathbf{B} = \mathbf{A}^{-1}$. In fact, since $\mathbf{B} = \mathbf{A}^T$, $\mathbf{B} = \mathbf{A}^{-1}$ means precisely that $\mathbf{A}^{-1} = \mathbf{A}^T$, which as you now know means that $\mathbf{A}$ is an orthogonal matrix!

7

Eigenvalues and Eigenvectors

This final chapter includes some of the most important and certainly the most advanced topics in this text. Eigenvectors and eigenvalues play a significant role in science and engineering. Many of these uses are too sophisticated to describe in a sophomore-level course, but nonethless, in this chapter we will cover two fundamental theorems of great practical value: The Spectral Theorem and Singular Value Decomposition. While the versions of these theorems stated here are for real matrices, actually proving them requires a small detour into the complex numbers. The good news is that the main general theorems that we have seen for real vector spaces also hold in complex vector spaces (or vector spaces with scalars in any field). Mainly we will have to make adjustments to the dot product to reflect the much richer structure of the complex numbers.

7.1   Eigenvalues and Eigenvectors

Suppose that $T : V \to V$ is a linear transformation of a vector space to itself. An *eigenvector* of T is a non-zero vector $\mathbf{v}$ such that $T(\mathbf{v}) = \lambda \mathbf{v}$ for some scalar λ. The scalar λ is called an *eigenvalue* of T and the eigenvalue of $\mathbf{v}$. That is, T takes $\mathbf{v}$ back into its own linear span. One possibility is that $\lambda = 0$, which means precisely that $\mathbf{v}$ is a non-zero vector in $\ker T$. First: not all matrices have eigenvectors. One of the simplest examples is 2×2 rotation matrices for almost all angles of rotation. This is easy enough to see geometrically—rotation by a non-zero angle rotates every non-zero vector at the origin, and only takes a vector of a scalar multiple of itself if the rotation is via an angle $2n\pi$ (which fixes every vector) or $2(n + 1)\pi$, which negates every vector. In the first case the eigenvalue is -1 and in the second case it is 1.

Another observation to make here is that there if there is one eigenvector then there are infinitely many. In fact, if $T(\mathbf{v}) = \lambda \mathbf{v}$ then for any scalar k,

$$T(k\mathbf{v}) = kT(\mathbf{v}) = k\,(\lambda \mathbf{v}) = \lambda(k\mathbf{v}).$$

That is, if $\mathbf{v}$ is an eigenvector with eigenvalue λ then so is $k\mathbf{v}$ for any scalar $k \neq 0$ ($k \neq 0$ is a necessary restriction because eigenvectors must be non-zero). Note that we *cannot* use a similar argument to show that the sum of eigenvectors is an eigenvector (which would show that eigenvectors are a subspace). The problem is that different eigenvectors may have different eigenvalues. So for example, if the eigenvalue for $\mathbf{v}$ is 2 and the eigenvalue for $\mathbf{w}$ is 4, then the argument breaks down:

$$T(\mathbf{v} + \mathbf{w}) = T(\mathbf{v}) + T(\mathbf{w}) = 2\mathbf{v} + 4\mathbf{w} \ldots$$

We cannot factor out the different numbers 2 and 4. The argument clearly *does work* if $\mathbf{v}$ and $\mathbf{w}$ have the same eigenvalue—something we will return to later.

Let's focus on $\mathbb{R}^n$ with its standard basis, which as you know means that all linear transformations from $\mathbb{R}^n$ to $\mathbb{R}^n$ are of the form $T_\mathbf{A}$ for some $n \times n$ matrix. In this case we will simply refer to the eigenvalues and eigenvectors of $\mathbf{A}$, rather than $T_\mathbf{A}$. That is, $\mathbf{v}$ is an eigenvalue of $\mathbf{A}$ if for some scalar λ,

$$\mathbf{Av} = \lambda \mathbf{v}$$

If $\mathbf{I}$ is the $n \times n$ identity matrix, then since $\mathbf{Iv} = \mathbf{v}$, the above equation can be rewritten as

$$\lambda \mathbf{Iv} - \mathbf{Av} = \mathbf{0}$$

which in turn is equivalent to

$$\left(\lambda \mathbf{I} - \mathbf{A} \right) \mathbf{v} = \mathbf{0}$$

Since $\mathbf{v} \neq \mathbf{0}$ this means that the matrix $\left(\lambda \mathbf{I} - \mathbf{A} \right)$ must be singular if $\mathbf{v}$ is an eigenvector, which means that $\det \left(\lambda \mathbf{I} - \mathbf{A} \right) = 0$. Therefore the eigenvalues of $\mathbf{A}$ (if there are any) are scalars λ that are solutions to the equation $\det \left(\lambda \mathbf{I} - \mathbf{A} \right) = 0$.

Definition 7.1 *The polynomial (with variable λ) given by $c(\lambda) := \det(\lambda \mathbf{I} - \mathbf{A})$ is called the characteristic polynomial of $\mathbf{A}$.*

Remark 7.1 *In the MPs and some examples to find eigenvalues we will use the polynomial $\det(\mathbf{A} - \lambda \mathbf{I})$ because it is a bit simpler to obtain $\mathbf{A} - \lambda \mathbf{I}$ from $\mathbf{A}$: you just have to subtract λ from the diagonal elements, rather than negating all the entries of $\mathbf{A}$ and adding λ to the diagonal elements. Why can we do this? As you know,*

$$\det \left(\lambda \mathbf{I} - \mathbf{A} \right) = \det \left((-1) \left(\mathbf{A} - \lambda \mathbf{I} \right) \right) \pm \det(\mathbf{A} - \lambda \mathbf{I})$$

(where the $\pm$ depends on whether the size of the matrix is even or odd). Since $\pm 0 = 0$, the characteristic polynomial has the same roots as $\det(\mathbf{A} - \lambda \mathbf{I})$, so we still get the eigenvalues when solving $\det(\mathbf{A} - \lambda \mathbf{I}) = 0$. You will see this in action in the next MP: you will simply subtract λ from the diagonal elements and then proceed with the determinant via cofactor expansion to find the characteristic polynomial.

Earlier we observed geometrically that rotations of angles between 0 and 2π other than $0, \pi, 2\pi$ have no eigenvalues. Let's check algebraically when eigenvalues exist and what they are if they do. As we saw earlier, the matrix representing rotation by angle θ is

$$\mathbf{R}_\theta = \begin{bmatrix} \cos\theta & \sin\theta \\ -\sin\theta & \cos\theta \end{bmatrix}$$

Then the eigenvalues λ must be solutions to

$$0 = \det\left(\begin{bmatrix} \lambda - \cos\theta & -\sin\theta \\ \sin\theta & \lambda - \cos\theta \end{bmatrix} \right) \tag{7.1}$$

$$= \lambda^2 - 2\lambda\cos\theta + \cos^2\theta + \sin^2\theta = \lambda^2 - 2\lambda\cos\theta + 1 \tag{7.2}$$

In other words, we seek the solutions to the above quadratic expression in the variable λ. As you will remember from high school, this quadratic equation has solutions if and only if the discriminant

$$4\cos^2\theta - 4 = 4(\cos^2\theta - 1) \geq 0$$

But $0 \leq \cos^2\theta \leq 1$, so there are no solutions except when $\cos^2\theta = 1$, which happens precisely when $\theta = 0, \pi, 2\pi$ (in the interval $[0, 2\pi]$). For $\theta = 0, 2\pi$, Equation (7.2) becomes

$$0 = \lambda^2 - 2\lambda + 1 = (\lambda - 1)^2$$

so the eigenvalue is 1, matching the fact that the rotation matrix by angle 0 or 2π is the identity matrix

$$\begin{bmatrix} 1 & 0 \\ 0 & 1 \end{bmatrix}$$

which multiplies all vectors by 1. Likewise, for $\theta = \pi$, for which the rotation matrix is

$$\begin{bmatrix} -1 & 0 \\ 0 & -1 \end{bmatrix}$$

Equation (7.2) becomes

$$0 = \lambda^2 + 2\lambda + 1 = (\lambda + 1)^2$$

In other words, we see that the rotation matrix in this case multiplies every vector by -1, and the eigenvalues are also -1.

In general, when $\mathbf{A}$ is a 2×2 matrix, finding eigenvalues is easy: $0 = \det\left(\begin{bmatrix} \lambda - a & -b \\ -c & \lambda - d \end{bmatrix} \right) = (\lambda - a)(\lambda - d) - bc$ is a quadratic equation, which you have known how to solve for years. For a 3×3 matrix this becomes

$$0 = \det \begin{bmatrix} \lambda - a & -b & -c \\ -d & \lambda - e & -f \\ -g & -h & \lambda - k \end{bmatrix}$$

You don't have to do the full cofactor expansion to realize that this equation is cubic. As you probably have heard, finding roots of polynomials gets ugly fast as the degree goes up. There are increasingly complex formulas for the roots of polynomials up to degree 4, and in the early 19th century it was shown that there cannot be an algebraic expression for the roots of polynomials of degree 5 and higher. Generally speaking for large matrices one must resort to numerical methods (i.e. using a computer) to approximate the roots (if they exist). In the next MP you will practice finding eigenvalues, and I promise you that the characteristic polynomials all do factor!

MP 71 *EIGENVALUES*

Once you have found eigenvalues, the task of finding the eigenvectors for those eigenvalues remains. For this, we simply look back at the definition. For any eigenvalue λ we seek all non-zero vectors $\mathbf{v}$ so that $(\mathbf{A} - \lambda\mathbf{I})\mathbf{v} = \mathbf{0}$ (as mentioned in Remark 7.1 this system has the same solutions as $(\lambda\mathbf{I} - \mathbf{A})\mathbf{v} = \mathbf{0}$). But you already know how to do this! You are simply solving a homogeneous system of equations. You can draw several conclusions from this. First, the set of eigenvectors for a particular eigenvalue λ, being the null space of a matrix, is a subspace of the domain. This space is called the *eigenspace* for the eigenvalue λ. By Theorem 6.15 you also know that the parameterized solutions that you obtain express the eigenvectors as a linear combination of a basis for the eigenspace. From this you can, in turn, determine the dimension of the eigenspace for the eigenvalue λ–which is precisely the nullity of $\mathbf{A} - \lambda\mathbf{I}$. In summary, you are using a process that you know well to obtain something new. You will practice this in the next MP.

MP 72 *EIGENVECTORS*

7.2   The Structure of the Characteristic Polynomial

We will next check that for an $n \times n$ matrix $\mathbf{A}$, the characteristic polynomial has degree n with leading coefficient 1. That is, c is of the form

$$c(\lambda) = \lambda^n + c_{n-1}\lambda^{n-1} + \cdots + c_1\lambda + c_0 \tag{7.3}$$

and at the same time we will figure out exactly what c_{n-1} and c_0 are. They are quantities that you have seen before! Actually, c_0 is easy. It is obtained by plugging $\lambda = 0$ into the polynomial, so we have

$$c_0 = c(0) = \det(0\mathbf{I} - \mathbf{A}) = \det(-\mathbf{A}) = (-1)^n \det \mathbf{A}$$

Now let's consider the case of a 2×2 matrix. Then

$$c(\lambda) = \det \begin{bmatrix} \lambda - a_{11} & a_{12} \\ a_{21} & \lambda - a_{22} \end{bmatrix} = (\lambda - a_{11})(\lambda - a_{22}) - a_{21}a_{12} \qquad (7.4)$$

$$= \lambda^2 - (a_{11} + a_{22})\lambda + (a_{11}a_{22} - a_{21}a_{12})$$

This is a quadratic with leading coefficient 1, as expected, and you should immediately recognize that $c_0 = \det \mathbf{A}$–which we showed already. As for c_1, you might not recognize it because you haven't seen it in a while: it is the negative of the sum of the diagonal elements, which is otherwise known as the trace $tr\mathbf{A}$. (I did promise you would see it again some day in an interesting place.) We will now show, by induction, that for any $n \times n$ matrix, c always has degree n with leading coefficient 1, and $c_{n-1} = -tr\mathbf{A}$. *Proof by induction* is closely related to the idea of an iterative definition, as we used, for example, for cofactor expansion. We have been informally using induction in various proofs already, so let's make it official. Since 1×1 matrices are not very interesting (although the statement is true for them as well, as you may check if you wish), I'll start with 2×2 matrices. We already proved this case above.

For the inductive step we suppose that we know the statement is true for any $(n-1) \times (n-1)$ matrix with $n-1 \geq 2$ (so our base case is included), and show that the statement is true for any $n \times n$ matrix. It then follows that the statement is true for any square matrix. Why? If it were not true, there would be some *smallest* matrix $\mathbf{B}$ (for "bad matrix") that doesn't obey this statement. Now $\mathbf{B}$ cannot be a 2×2 matrix, because we proved that case. So $\mathbf{B}$ is an $n \times n$ matrix for some $n > 2$. But remember, $\mathbf{B}$ was the *smallest such matrix*, so the statement is true for any $(n-1) \times (n-1)$ matrix. But our inductive step says that if the statement is true for any $(n-1) \times (n-1)$ then it is true for any $n \times n$ matrix, and therefore must be true for $\mathbf{B}$, which is a contradiction. This same argument justifies inductive proofs in general.

Returning to the inductive step, we are assuming that the statement is true for any $(n-1) \times (n-1)$ matrix, with $n-1 \geq 2$, and showing that the statement is true for any $n \times n$ matrix $\mathbf{A}$. The characteristic polynomial for $\mathbf{A}$ is

$$p(\lambda) = \det \begin{bmatrix} \lambda - b_{11} & b_{12} & \cdots & b_{1n} \\ b_{21} & \lambda - b_{22} & \cdots & b_{2n} \\ \vdots & \vdots & \vdots & \vdots \\ b_{n1} & \cdots & \cdots & \lambda - b_{nn} \end{bmatrix} \qquad (7.5)$$

Doing cofactor expansion across the top row, the only entry with a "λ" in it is the first, and it is multiplied by the determinant of the minor matrix $\mathbf{A}_{11}$, which is precisely of the form $\lambda\mathbf{I} - \mathbf{A}_{11}$. That is, the first term in the expansion is $(\lambda - a_{11})(\det(\lambda\mathbf{I} - \mathbf{A}_{11}))$, and $\det(\lambda\mathbf{I} - \mathbf{A}_{11})$ is precisely the characteristic polynomial of the $(n-1) \times (n-1)$ matrix $\mathbf{A}_{11}$. Note that $tr\mathbf{A} = b_{11} + b_{22} + \cdots + b_{nn} = b_{11} + tr\mathbf{A}_{11}$. Now we know by the inductive step that the characteristic polynomial of $\mathbf{A}_{11}$ is of the form $\lambda^{n-1} - tr\mathbf{A}_{11}\lambda^{n-2} + \cdots + d_0$.

Putting all this together, the first term in the cofactor expansion is

$$(\lambda - b_{11})(\lambda^{n-1} + -tr\mathbf{A}_{11}\lambda^{n-2} + \cdots + d_0) = \lambda^n - (b_{11} + \lambda r\mathbf{A}_{11})\lambda^{n-1} + \cdots$$

$$= \lambda^n - tr\mathbf{A}\lambda^{n-1} + \cdots$$

Which is what we wanted to show. But what about the rest of the terms in the cofactor expansion across the top row? I'll leave it to you to explain that the other terms do not contain any λ^n or λ^{n-1} terms. This finishes the proof that $c_{n-1} = -tr\mathbf{A}$, finishing the inductive step. We formally state the theorem:

Theorem 7.1 *The characteristic polynomial* $\det(\lambda\mathbf{I} - \mathbf{A})$ *of an* $n \times n$ *matrix* **A** *is of the form*

$$c(\lambda) = \lambda^n - tr\mathbf{A}\lambda^{n-1} + \cdots + c_1\lambda + (-1)^n \det \mathbf{A}$$

Exercise 7.1 *Explain in your own words why the remaining terms in the cofactor expansion of (7.5) do not contain any* λ^{n-1} *terms. Hint: the only* t's *are on the diagonal.*

Example 7.1 *Let's look at one of the first concrete matrices that you saw in this course (1.5):* $\mathbf{M}_1 = \begin{bmatrix} 1 & 0 & 0 \\ 0 & 0 & -1 \\ 0 & 1 & 0 \end{bmatrix}$. *Earlier you determined geometrically that this matrix represents rotation about the x-axis. Geometrically speaking this means that there should be only one eigenvalue, 1, and all eigenvectors must lie on the x-axis. To check this more precisely, use the fact that the columns of* $\mathbf{M}_1$ *are the images of the standard basis of* $T_{\mathbf{M}_1}$, *from which you can now simply read off what you initially had to compute:*

$$T_{\mathbf{M}_1}(1,0,0) = (1,0,0) \quad T_{\mathbf{M}_1}(0,1,0) = (0,0,1) \quad T_{\mathbf{M}_1}(0,0,1) = (0,-1,0)$$

You immediately see that $\mathbf{e}_1$ *is an eigenvector with eigenvalue 1, as expected geometrically. Let's now compute the eigenvalues and eigenvectors of* $\mathbf{M}_1$ *algebraically to verify this. This will also verify the fact that, as is geometrically evident, there are no other eigenvalues or eigenvectors not parallel to* $\mathbf{e}_1$. *Using the strategy of Remark 7.1 and cofactor expansion,*

$$\det(\mathbf{M}_1 - \lambda\mathbf{I}) = \det \begin{bmatrix} 1-\lambda & 0 & 0 \\ 0 & -\lambda & -1 \\ 0 & 1 & -\lambda \end{bmatrix} = (1-\lambda)(\lambda^2 + 1)$$

The characteristic polynomial is already factored, and you can immediately see that there is only one real root: $\lambda = 1$. *To find the eigenvector(s) for this eigenvalue, we replace* λ *by 1 in the above matrix to obtain*

$$\begin{bmatrix} 0 & 0 & 0 \\ 0 & -1 & -1 \\ 0 & 1 & -1 \end{bmatrix}$$

Augmenting with **0** *and using row operations, we obtain*

$$
\begin{bmatrix}
0 & 1 & -1 & | & 0 \\
0 & -1 & -1 & | & 0 \\
0 & 0 & 0 & | & 0
\end{bmatrix}
\rightarrow
\begin{bmatrix}
0 & 1 & -1 & | & 0 \\
0 & 0 & -2 & | & 0 \\
0 & 0 & 0 & | & 0
\end{bmatrix}
$$

Without going any further we can solve this system: $x = t$, $z = 0, y = 0$. *That is, the eigenvectors are of the form*

$$
\begin{bmatrix} t \\ 0 \\ 0 \end{bmatrix} = t\mathbf{e}_1
$$

as expected.

Remark 7.2 *It can be shown that any 3×3 matrix with positive entries has a positive real eigenvalue. This theorem is known as Frobenious's Theorem, and there are related theorems for dimensions greater than 3. The proof of this very useful theorem is quite deep. For example, a standard proof uses an area known as algebraic topology, typically learned in graduate mathematics courses. Algebraic topology has many deep applications to modern science, including an area known as topological data analysis, which can be used to analyze large data sets that are less suited to analysis using more old fashioned techniques.*

Exercise 7.2 *For this exercise we will spice up the matrix $\mathbf{M}_2$ to allow rotations about the z-axis of any angle. That is*

$$
\mathbf{M}_\theta =
\begin{bmatrix}
\cos\theta & -\sin\theta & 0 \\
\sin\theta & \cos\theta & 0 \\
0 & 0 & 1
\end{bmatrix}
$$

1. Explain why you can read immediately from the columns of the matrix the matrix that $\mathbf{e}_3$ is an eigenvector with eigenvalue 1, meaning that the z-axis is fixed, and the matrix rotates the (x, y)-plane by angle θ.

2. Directly calculate the eigenvalues and eigenvectors of $\mathbf{M}_\theta$, as functions of θ. Hint: use cofactor expansion in the bottom row, being careful to write down the correct sign for the only non-zero term.

3. Analyze the meaning of the eigenvalues along the lines of the discussion above in this section involving rotation matrices.

Exercise 7.3 *Show that a matrix $\mathbf{A}$ and its transpose have the same eigenvalues. Hint: The determinant of the transpose of a matrix is equal to the determinant of the matrix.*

Exercise 7.4 *Show that if* $\mathbf{A}$ *is non-singular then the eigenvalues of* $\mathbf{A}^{-1}$ *(if it has them) are* $\frac{1}{\lambda}$ *for every eigenvalue* λ *of* $\mathbf{A}$*, and the eigenvectors of* $\frac{1}{\lambda}$ *for* $\mathbf{A}^{-1}$ *are the same as the eigenvectors of* λ *for* $\mathbf{A}$*. Hint: First, explain why* $\lambda = 0$ *is impossible in this case. Then multiply* $\mathbf{Av} = \lambda \mathbf{v}$ *on the left on both sides by* $\mathbf{A}^{-1}$*.*

Exercise 7.5 *Use the above exercises to show that if* $\mathbf{A}$ *is orthogonal then the eigenvalues of* $\mathbf{A}$ *(if they exist) are* ± 1*. This makes sense because as was worked out earlier, orthogonal matrices preserve the lengths of vectors.*

Exercise 7.6 *Show directly that the eigenspace of a specific eigenvalue* λ *is a subspace of the domain. That is, suppose that* $\mathbf{v}$ *and* $\mathbf{w}$ *are eigenvectors for eigenvalue* λ *for a matrix* $\mathbf{A}$ *and* k *is a scalar. Check that* $k\mathbf{v} + \mathbf{w}$ *is also an eigenvector with eigenvalue* λ*.*

Remark 7.3 *I will now provide a warning: Gauss-Jordan elimination preserves many things, including solutions of systems, rank, nullity, etc. It is used to find eigenvalues and eigenvectors (but not with the original matrix). However, Gauss-Jordan does not preserve eigenvalues or eigenvectors! This is completely obvious if you think about the fact that every non-singular square matrix (even ones without eigenvalues) can be reduced to* $\mathbf{I}$ *with row operations, and every vector is an eigenvector of* $\mathbf{I}$ *with eigenvalue 1. This makes the next theorem a little less useful than it may seem at first.*

Theorem 7.2 *If* $\mathbf{A}$ *is a an upper (or lower) triangular matrix then its diagonal elements are its eigenvalues.*

Proof. As you know, the determinant of a triangular matrix is the product of its diagonal elements, so

$$
\det \begin{bmatrix} a_{11} - \lambda & a_{12} & \cdots & a_{1n} \\ 0 & a_{22} - \lambda & \cdots & a_{2n} \\ \vdots & \vdots & \vdots & \vdots \\ 0 & 0 & 0 & a_{nn} - \lambda \end{bmatrix} = (a_{11} - \lambda)(a_{22} - \lambda)\cdots(a_{nn} - \lambda)
$$

The characteristic polynomial is already factored for you! ∎

But to reiterate, you *cannot* find the eigenvalues of a matrix by simply using Gauss-Jordan Elimination to reduce it to an upper triangular matrix–the diagonals in general are not eigenvalues of the original matrix.

Theorem 7.3 *The intersection of eigenspaces corresponding to different eigenvalues of a matrix is* $\{\mathbf{0}\}$*. Put another way, an eigenvector* $\mathbf{v}$ *cannot be in more than one eigenspace.*

Proof. This theorem is simply a consequence of the definition. If $\mathbf{v}$ is an eigenvector with eigenvalue $\mathbf{v}$, this means that $\mathbf{Av} = \lambda \mathbf{v}$. Since $\mathbf{v} \neq \mathbf{0}$, it is impossible for $\mathbf{Av} \neq \lambda' \mathbf{v}$ for any $\lambda' \neq \lambda$. ∎

It may be tempting at this point, given Theorem 6.22, to speculate that the eigenspaces for different eigenvalues are orthogonal. The next exercise shows that this is not the case in general, but will see this behavior in a the extremely important case of symmetric matrices (Theorem 7.10).

Exercise 7.7 *Show that* $\begin{bmatrix} 1 & 0 & 0 \\ 0 & 1 & 1 \\ 0 & 0 & -1 \end{bmatrix}$ *has two eigenvalues, ± 1, and that the eigenvectors for -1 are not orthogonal to some eigenvectors for 1. Note that this matrix is not symmetric.*

7.3 Diagonalization

Definition 7.2 *A matrix $\mathbf{A}$ is called diagonalizable if it is similar to a diagonal matrix. That is, there is there an invertible matrix $\mathbf{P}$ such that $\mathbf{P}^{-1}\mathbf{A}\mathbf{P}$ is a diagonal matrix $\mathbf{D}$.*

You have already diagonalized some matrices in MP 67. In that case, you were simply given matrices that were known (by me!) to be diagonalizable. But in general, how do you determine whether a matrix is diagonalizable, and if so, how do we find the matrices $\mathbf{P}$ and $\mathbf{D}$? The next theorem completely answers this question.

Theorem 7.4 *An $n \times n$ matrix $\mathbf{A}$ is diagonalizable if and only if $\mathbb{R}^n$ has a basis $B = \{\mathbf{v}_1, \ldots, \mathbf{v}_n\}$ consisting of eigenvectors of $\mathbf{A}$. If $\mathbf{D}$ is the diagonalization of $\mathbf{A}$ then*

$$\mathbf{D} = \begin{bmatrix} \lambda_1 & \cdots & 0 \\ \vdots & \ddots & \vdots \\ 0 & \cdots & \lambda_n \end{bmatrix} \tag{7.6}$$

where λ_i is the eigenvalue of $\mathbf{v}_i$. Moreover, $\mathbf{D} = \mathbf{P}^{-1}\mathbf{A}\mathbf{P}$, where $\mathbf{P} = \mathbf{C}_B$.

Proof. Suppose that $\mathbb{R}^n$ has a basis $B = \{\mathbf{v}_1, \ldots, \mathbf{v}_n\}$ of eigenvectors of $\mathbf{A}$, with eigenvalues $\lambda_1, \ldots, \lambda_n$. Then by Theorem 6.9, $\mathbf{A}$ is similar to the matrix $\mathbf{D}$ that represents $T_\mathbf{A}$ with respect to this basis. To find the ith column of $\mathbf{D}$ we express $T_\mathbf{A}(\mathbf{v}_i)$ as a linear combination of the vectors in B and take the coefficients. Well

$$T_\mathbf{A}(\mathbf{v}_i) = \lambda_i \mathbf{v}_i = 0\mathbf{v}_1 + \cdots + \lambda_i \mathbf{v}_i + \cdots + 0\mathbf{v}_n$$

and therefore the ith column of $\mathbf{D}$ is

$$\begin{bmatrix} 0 \\ \vdots \\ 0 \\ \lambda_i \\ 0 \\ \vdots \\ 0 \end{bmatrix}$$

where the λ_i is in the ith row. This finishes the "if" statement. This logical implication is also known as "sufficiency"; that is, for $\mathbf{A}$ to be diagonalizable it is *sufficient* that $\mathbb{R}^n$ have a basis of eigenvalues of $\mathbf{A}$.

The converse statement is "necessity"; that is, if $\mathbf{A}$ is diagonalizable it is *necessary* that $\mathbb{R}^n$ have a basis of eigenvalues of $\mathbf{A}$. To prove this implication, we suppose that $\mathbf{A} = \mathbf{PDP}^{-1}$ for some diagonal matrix $\mathbf{D}$. Then $T_\mathbf{A}(\mathbf{P}) = \mathbf{AP} = \mathbf{PD}$. Now $T_\mathbf{A}(\mathbf{P}^i)$ is the ith column of $\mathbf{AP}$, which is the ith column of $\mathbf{PD}$. Then

$$T_\mathbf{A}(\mathbf{P}) = \mathbf{PD} = [\mathbf{P}^1 \mid \cdots \mid \mathbf{P}^n] \begin{bmatrix} \lambda_1 & \cdots & 0 \\ \vdots & \ddots & \vdots \\ 0 & \cdots & \lambda_n \end{bmatrix} = [\lambda_1 \mathbf{P}^1 \mid \cdots \mid \lambda_n \mathbf{P}^n]$$

That is, $T_\mathbf{A}(\mathbf{P}^i) = \lambda_i \mathbf{P}^i$, so the columns of $\mathbf{P}$ are a basis of eigenvectors of $T_\mathbf{A}$. ∎

Theorem 7.5 *If* $S = \{\mathbf{v}_1, \ldots, \mathbf{v}_k\}$ *are eigenvectors with eigenvalues* $\{\lambda_1, \ldots, \lambda_k\}$ *and* $\mathbf{v}$ *is an eigenvector with eigenvalue* $\lambda \neq \lambda_i$ *for all* i, *then* $\mathbf{v}$ *is not in the span of* S. *In particular, if* S *is linearly independent then the set* $\{\mathbf{v}_1, \ldots, \mathbf{v}_k, \mathbf{v}\}$ *is linearly independent.*

Proof. First suppose that the vectors in S are linearly independent. Suppose, to the contrary, that $\mathbf{v} = \sum_{\iota=1}^k c_i \mathbf{v}_i$. Applying $T_\mathbf{A}$ to each side of this equation gives us

$$\lambda \mathbf{v} = T_\mathbf{A}\left(\sum_{\iota=1}^k c_i \mathbf{v}_i\right) = \sum_{\iota=1}^k c_i T_\mathbf{A}(\mathbf{v}_i) = \sum_{\iota=1}^k c_i \lambda_i \mathbf{v}_i$$

Therefore

$$\sum_{\iota=1}^k c_i \lambda_i \mathbf{v}_i - \lambda \mathbf{v} = 0$$

Now it is impossible for $\lambda = 0$ because this would contradict our assumption that S is linearly independent. But since $\lambda \neq 0$ we can divide by it. This gives us

$$\sum_{\iota=1}^k c_i \frac{\lambda_i}{\lambda} \mathbf{v}_i = \mathbf{v}$$

which contradicts the assumption that $\mathbf{v}$ is not in the span of S.

Now if S is not linearly independent, by the Spanning Set Reduction Theorem (Theorem 6.6) we can find a linearly independent subset of S that has the same span. This reduces us to the above case and finishes the proof that $\mathbf{v}$ is not in the span of S.

For the last statement of the theorem, note that if S is linearly independent, then since $\mathbf{v}$ is not in the span of S, by the Basis Extension Theorem (Theorem 6.4), $\{\mathbf{v}_1, \ldots, \mathbf{v}_k, \mathbf{v}\}$ is linearly independent. ∎

The next theorem is quite useful because in some cases it allows you to avoid the problem of checking the dimension of eigenspaces to verify diagonalizability.

Theorem 7.6 *Suppose* $\lambda_1, \ldots, \lambda_k$ *are distinct (i.e. no two are equal) eigenvalues of an* $n \times n$ *matrix* $\mathbf{A}$*, with corresponding eigenvectors* $S = \{\mathbf{v}_1, \ldots, \mathbf{v}_k\}$*. Then* S *is linearly independent. In particular, if* $\mathbf{A}$ *has* n *distinct eigenvalues then* $\mathbf{A}$ *is diagonalizable.*

Proof. Let's iteratively show that $\{\mathbf{v}_1, \ldots, \mathbf{v}_k\}$ is linearly independent. Since $\mathbf{v}_1 \neq \mathbf{0}$, $\{\mathbf{v}_1\}$ is linearly independent. Since $\lambda_1 \neq \lambda_2$, by Theorem 7.5 $\{\mathbf{v}_1, \mathbf{v}_2\}$ is linearly independent. Likewise, since $\lambda_3 \neq \lambda_1$ and $\lambda_3 \neq \lambda_2$, $\{\mathbf{v}_1, \mathbf{v}_2, \mathbf{v}_3\}$ is linearly independent, and we may continue this process to all the vectors in S. ∎

Let's enumerate the possibilities for a 3×3 matrix $\mathbf{A}$. Since the characteristic polynomial of $\mathbf{A}$ has degree 3, $\mathbf{A}$ has between 0 and 3 distinct eigenvalues. If there aren't any eigenvalues then there is nothing more to say about them.

For each possibility from 1-3 eigenvalues we have the following scenarios:

1. When $\mathbf{A}$ has a single eigenvalue λ, λ may have an eigenspace of dimension 1, 2, or 3. Then $\mathbf{A}$ is diagonalizable if and only if the dimension of the eigenspace is 3, meaning that $\mathbf{A}$ has a basis of eigenvectors. However, as is shown in the exercise below, in this case $\mathbf{A} = \lambda \mathbf{I}$, so $\mathbf{A}$ is already diagonal!

2. Suppose that $\mathbf{A}$ has two eigenvalues $\lambda_1 \neq \lambda_2$. I claim that the *sum* of the dimensions of the eigenspaces is at most 3. If the eigenspace for λ_1 had a basis $\{\mathbf{v}_1, \ldots, \mathbf{v}_3\}$ and $\mathbf{v}$ were an eigenvector for eigenvalue λ_2, then by Theorem 7.5, $\{\mathbf{v}_1, \mathbf{v}_2, \mathbf{v}_3, \mathbf{v}\}$ would be linearly independent, which is impossible in $\mathbb{R}^3$. The only other possibility is if the eigenspaces for λ_1 and λ_2 are each of dimension 2. Suppose $\{\mathbf{v}_1, \mathbf{v}_2\}$ is a basis for the first eigenspace and $\{\mathbf{w}_1, \mathbf{w}_2\}$ is a basis for the second. We will reach a contradiction by showing that the set $\{\mathbf{v}_1, \mathbf{v}_2, \mathbf{w}_1, \mathbf{w}_2\}$ is linearly independent, which is impossible in $\mathbb{R}^3$. Suppose we write $c_1\mathbf{v}_1 + c_2\mathbf{v}_2 + c_3\mathbf{w}_1 + c_4\mathbf{w}_2 = \mathbf{0}$, we will get our contradiction by showing that it must be true that $c_i = 0$ for all i. We can write

$$c_1\mathbf{v}_1 + c_2\mathbf{v}_2 = -c_3\mathbf{w}_1 - c_4\mathbf{w}_2$$

This means that $c_1\mathbf{v}_1 + c_2\mathbf{v}_2$ is also in the eigenspace for λ_2, which is impossible by Theorem 7.3 unless $c_1 = c_2 = 0$. Likewise $c_3 = c_4 = 0$, finishing the proof.

We now know that since eigenspaces have a minimum dimension of 1 and a maximum dimension of 2, $\mathbf{A}$ is diagonalizable if and only if one of the eigenspaces has dimension 2. This leaves one more possibility:

3. When $\mathbf{A}$ has three distinct eigenvalues then Theorem 7.6 tells you that $\mathbf{A}$ is diagonalizable.

Remark 7.4 *The above discussion can be extended to higher dimensions, but there are more efficient arguments beyond the scope of this course.*

Exercise 7.8 *Show that if* $\mathbf{Av} = \lambda\mathbf{v}$ *for every vector* $\mathbf{v}$ *then* $\mathbf{A} = \lambda\mathbf{I}$. *Hint: As you know by now, the ith column of* $\mathbf{A}$ *is* $\mathbf{Ae}_i$.

MP 73 *DIAGONALIZATION*

MP 74 *DIAGONALIZATION II*

Now let's consider the relationship between the dimension of an eigenspace and the power of the corresponding linear factor in the characteristic polynomial. We will look at two of the examples that you did in MP 73. Consider first

$$\begin{bmatrix} 5 & 8 & 16 \\ 4 & 1 & 8 \\ -4 & -4 & -11 \end{bmatrix}$$

which had characteristic polynomial $(\lambda + 3)^2(\lambda - 1)$ and had the following bases for its eigenspaces (choosing specific values of the parameters in the parameterized solutions):

$$\left\{ \begin{bmatrix} -2 \\ -1 \\ 1 \end{bmatrix} \right\} \text{ for eigenvalue } 1$$

and

$$\left\{ \begin{bmatrix} -1 \\ 1 \\ 0 \end{bmatrix}, \begin{bmatrix} -2 \\ 0 \\ 1 \end{bmatrix} \right\} \text{ for eigenvalue } -3$$

You should notice that the dimension of each eigenspaces is equal to the power of the corresponding linear factor in the characteristic polynomial. That is, the linear factor corresponding to 1 is $(\lambda - 1) = (\lambda - 1)^1$, and the dimension of the eigenspace is 1. Likewise the linear factor corresponding to -3 is $(\lambda + 3)^2$, and the dimension of the eigenspace is 2.

But now consider the matrix

$$\begin{bmatrix} 1 & -2 & 3 \\ 2 & 6 & -6 \\ 1 & 2 & -1 \end{bmatrix}$$

which has characteristic polynomial $(\lambda - 2)^3$ and basis eigenvectors

$$\left\{ \begin{bmatrix} -2 \\ 1 \\ 0 \end{bmatrix}, \begin{bmatrix} 3 \\ 0 \\ 1 \end{bmatrix} \right\}$$

for its one and only eigenspace. In this case the power of the corresponding linear factor $(\lambda - 2)$ in the characteristic polynomial is 3. But the dimension of the eigenspace is 2. The next general theorem has a proof that is beyond the scope of this text:

Theorem 7.7 *If* **A** *is a square matrix with characteristic polynomial that includes linear factors*

$$(\lambda - e_1)^{n_1}(\lambda - e_2)^{n_2} \cdots (\lambda - e_m)^{n_m} \tag{7.7}$$

then the dimension of the eigenspace corresponding to the eigenvalue e_i is at most n_i.

Example 7.2 *If the characteristic polynomial of* **A** *factors as $(\lambda - 1)^2(\lambda + 1)^3(\lambda - 6)$, the dimension of the eigenspace for eigenvalue 1 is at most 2, for eigenvalue -1 the dimension is at most 3, and for eigenvalue 6 it is equal to 1 (since it must be at least 1).*

Do not forget the "at most" part–generally these two numbers are not equal. Sometimes the corresponding power n_i is called the "algebraic multiplicity" and the dimension of the eigenspace for e_i is the "geometric multiplicity".

Exercise 7.9 *Show that a square matrix* **A** *is diagonalizable if and only if the characteristic polynomial $c(\lambda)$ completely factors into linear factors as*

$$c(\lambda) = (\lambda - e_1)^{n_1}(\lambda - e_2)^{n_2} \cdots (\lambda - e_m)^{n_m}$$

and for each eigenvalue, the geometric multiplicity is equal to the algebraic multiplicity. Hint: This is not as hard as it may seem at first glance. Really this amounts to counting the dimensions of the eigenspaces and applying Theorem 7.4.

Diagonalizability makes many things easier. Many of the advantages of diagonalizable matrices are beyond the level of this text, but there is one simple example that shows its utility: powers of matrices. You may recall that finding powers of matrices is important to understand processes governed by stochastic matrices. Suppose that you have a matrix $\mathbf{A}$, and you need to compute the product of $\mathbf{A}$ with itself 1000 times; that is, you need to compute the matrix $\mathbf{A}^{1000}$. If $\mathbf{A}$ is a large matrix, that's a lot of computation. But this is much easier to do with a diagonal matrix because matrix multiplication is simply multiplication of the diagonal elements. For example,

$$\begin{bmatrix} 2 & 0 & 0 & 0 \\ 0 & -1 & 0 & 0 \\ 0 & 0 & 0 & 0 \\ 0 & 0 & 0 & 4 \end{bmatrix}^{1000} = \begin{bmatrix} 2^{1000} & 0 & 0 & 0 \\ 0 & (-1)^{1000} & 0 & 0 \\ 0 & 0 & 0 & 0 \\ 0 & 0 & 0 & 4^{1000} \end{bmatrix}$$

Now suppose that $\mathbf{A}$ is not necessarily diagonal, but is diagonalizable. That is $\mathbf{D} = \mathbf{P}^{-1}\mathbf{A}\mathbf{P}$ is a diagonal matrix. But then $\mathbf{A} = \mathbf{P}\mathbf{D}\mathbf{P}^{-1}$, and since $\mathbf{P}^{-1}\mathbf{P} = \mathbf{I}$,

$$\mathbf{A}^2 = \left(\mathbf{P}\mathbf{D}\mathbf{P}^{-1}\right)^2 = \mathbf{P}\mathbf{D}\mathbf{P}^{-1}\mathbf{P}\mathbf{D}\mathbf{P}^{-1} = \mathbf{P}\mathbf{D}\mathbf{I}\mathbf{D}\mathbf{P}^{-1} = \mathbf{P}\mathbf{D}^2\mathbf{P}^{-1}$$

Likewise, $\mathbf{A}^3 = \mathbf{P}\mathbf{D}^3\mathbf{P}^{-1}$ and so on. In fact

$$\mathbf{A}^{1000} = \mathbf{P}\mathbf{D}^{1000}\mathbf{P}^{-1}$$

This means that computing high powers of a diagonalizable matrix takes almost no more effort than computing the same high power of a diagonal matrix. So the relatively small effort of determining the matrix $\mathbf{P}$ pays off very well for computing large powers of $\mathbf{A}$.

Exercise 7.10 *Prove that if λ is an eigenvalue of a matrix $\mathbf{A}$ with eigenvector $\mathbf{v}$, then λ^k is an eigenvalue of $\mathbf{A}^k$ with eigenvector $\mathbf{v}$. Hint: you may either prove it for general k using appropriate "$\cdots$"s or write it out fully for $k = 3$, in which case your argument should start with $\mathbf{A}^3\mathbf{v} = \mathbf{A}\mathbf{A}\mathbf{A}(\mathbf{v}) = \cdots$.*

7.4   Equivalent Conditions to be Invertible

Here is the final and somewhat long list of conditions that are equivalent to an $n \times n$ matrix $\mathbf{A}$ being invertible (non-singular). All of these statements have been proved at various points in this text–it may be a useful exercise for you to piece it all together. There are further equivalent conditions that can be derived from these and other theorems. For example, since there are n columns (and rows) in $\mathbf{A}$, then in Condition 8 (and 9), "linearly independent" may be replaced by "a spanning set" or "a basis".

1. The reduced echelon form of $\mathbf{A}$ is $\mathbf{I}$.

2. $\mathbf{A}$ is a product of elementary matrices.

3. $\mathbf{AX} = \mathbf{0}$ has only the trivial solution.

4. $\mathbf{AX} = \mathbf{B}$ is consistent for every vector $\mathbf{B}$.

5. $\mathbf{AX} = \mathbf{B}$ has a unique solution for every vector $\mathbf{B}$.

6. $\det \mathbf{A} \neq 0$

7. $\mathbf{A}^T$ is invertible.

8. The columns $\{\mathbf{A}^1, \ldots, \mathbf{A}^n\}$ are linearly independent.

9. The rows $\{\mathbf{A}_1, \ldots, \mathbf{A}_n\}$ are linearly independent.

10. The rank of $\mathbf{A}$ is n (sometimes this is called "full rank").

11. Then nullity of $\mathbf{A}$ is 0.

12. 0 is not an eigenvalue of $\mathbf{A}$.

7.5 Complex Numbers with Applications to Eigenvalues

The *grand finale* of this text will be Singular Value Decomposition (SVD), an important theorem used in many modern problems in science and engineering. In order to understand and prove this theorem, we will need to introduce (at long last) the field of complex numbers $\mathbb{C}$. $\mathbb{C}$ has an extraordinarily rich geometric and algebraic structure, and we will only touch on its basic properties here. The principal reason to introduce $\mathbb{C}$ at this point is as follows. As you will recall, eigenvalues are the roots of the characteristic polynomial, and not all polynomials with real coefficients have roots in the real numbers. For example, according to the calculation (7.2), the characteristic polynomial of the matrix $\mathbf{R}_{\frac{\pi}{2}}$ of rotation by angle $\frac{\pi}{2}$ is $\lambda^2 + 1$, which has no real roots because it is always positive. But a theorem sometimes called the "Fundamental Theorem of Algebra" says that every polynomial *does* have roots in $\mathbb{C}$. This means that every matrix, real or complex, has a (possibly complex) eigenvalue! Returning to $\mathbf{R}_{\frac{\theta}{2}}$, the complex root of $\lambda^2 + 1$ is the complex number i, which satisfies $i^2 = -1$. The strategy to construct $\mathbb{C}$ is to extend $\mathbb{R}$ to a bigger field that contains i. More precisely, the complex numbers $\mathbb{C}$ consist of all numbers z of the form $a + bi$, where a and b are real numbers, and i satisfies $i^2 = -1$. Complex numbers are added and multiplied as you might expect, i.e.

$$(a + bi) + (c + di) = (a + c) + (b + d)i$$

and, "FOILING" as you might have done in high school and using $i^2 = -1$,

$$(a + bi)(c + di) = (ac - bd) + (ad + bc)i$$

The real numbers $\mathbb{R}$ are a subset of $\mathbb{C}$ consisting of the complex numbers of the form $a + 0i$, which we normally simply write as a even when considering them as complex numbers. In fact $\mathbb{R}$ is a "subfield" of $\mathbb{C}$ in the sense that it is a field with the same operations as $\mathbb{C}$. The 0 of $\mathbb{C}$ is just the real number $0 = 0 + 0i$. The 1 of $\mathbb{C}$ is just the real number $1 = 1 + 0i$. It is not hard to check that $\mathbb{C}$ is a field with these definitions, although multiplicative inverses require a little linear algebra! See Exercise 7.12. The "Fundamental Theorem of Algebra" is now simple enough to state, but it took mathematicians more than a century to prove it: every complex polynomial $a_n z^n + \cdots + a_1 z + a_0$, i.e. the variable z and coefficients a_i are complex numbers, has a root in $\mathbb{C}$. This extends the idea that i and $-i$ are the roots of $z^2 + 1$. An immediate consequence is that every complex matrix has eigenvalues. Since real matrices can also be considered as complex matrices, real matrices also have (possibly complex) eigenvalues.

Exercise 7.11 *Add and multiply the following pairs of complex numbers:* $z = 3 + 2i$ *and* $w = -1 - i$.

Exercise 7.12 *Find* $\frac{1}{a+bi}$, *assuming* a *or* b *is non-zero. Hint: you seek a complex number* $c + di$ *such that* $(a + bi)(c + di) = 1 = 1 + 0i$. *FOILING the left side gives you a complex number* $e + if$. *Setting* $e = 1$ *and* $f = 0$ *gives you two linear equations with variables* c, d, *which are the real numbers you are trying to find.*

Exercise 7.13 *Find the multiplicateive inverse of* $-2i$.

The letter "i" stands for "imaginary", a term that goes back to Descartes, who wanted to distinguish numbers like i from "real" numbers that measured physical quantities. To this day there are silly debates on the internet about whether complex numbers are "real". Like many endless pseudo-math arguments, this one centers on the meaning of a word in popular discourse, not its mathematical definition. Of course no one wants to admit this, because that would ruin the fun! (Assuming it *is* fun.) In this case the word that nobody clearly defines is "real"—does that mean it exists (what does that mean?), and if it exists does it exist as an abstract thing or a concrete thing? Blah, blah, blah. None of that is of the slightest practical value. In mathematics, we define our words carefully and our conclusions are logically based on those definitions. In this case, since $x^2 + 1$ has no real roots, and i is a root of this polynomial, i is, according to our definition, *not real*! End of debate.

The next thing we need to do is extend the notion of dot product to the complex numbers. Since $\mathbb{C}$ is a field, we can consider the complex vector space $\mathbb{C}^n$ consisting of n-tuples $(z_1, \ldots, z_n)$. As with $\mathbb{R}^n$, one can define concepts such as subspaces, linear independence, bases, dimension, etc. Likewise an n-dimensional complex vector space may be abstractly defined using the same definitions as for real vector spaces, but with complex scalars. By the same arguments as in the real case, each n-dimensional complex vector space is linerly isomorphic to $\mathbb{C}^n$. What about the dot product? As you have seen in the case of defining the product of matrices, sometimes the most naive idea is not the most useful one. Let's start with the vector space $\mathbb{C}$, which we can think of as the plane. In fact, every complex number $a+bi$ corresponds to exactly one point (a, b) in the plane, and vice-versa. The plane has a much richer geometry than the line, and this needs to be considered when deciding how to extend the dot product to $\mathbb{C}^n$. By the way, since $\mathbb{C}$ is geometrically the plane, does that mean it is one-dimensional or two-dimensional? We have definitions for that! Considered as a real vector space, the plane is two-dimensional because it has a basis with two vectors, e.g. $\{\mathbf{e}_1, \mathbf{e}_2\}$. But as a complex vector space, $\mathbb{C}$ has a basis consisting of the complex number 1 (or any non-zero complex number for that matter) because every complex number z is a (complex) scalar multiple of 1. So the answer is that according to our definitions, $\mathbb{C}$ has dimension 1 when considered as complex vector space.

In $\mathbb{R}$, $|x|$ is the way in which we measure the "magnitude" of x. Put another way, the non-negative number $|x|$ describes "how far away x is from 0". For a complex number $a + bi$, which we can consider as the point (a, b) in the plane, by the Pythagorean Theorem the distance from the origin is $\sqrt{a^2 + b^2}$. So that's a natural measure of "magnitude" of $a + b_i$. Now how do we get the real number $a^2 + b^2$ from the complex number $a + bi$?

Exercise 7.14 *Show that* $(a + bi)(a - bi) = a^2 + b^2$.

Definition 7.3 *For* $z = a + bi$ *the number* $a - bi$ *is clearly important, and so it is given a name, the complex conjugate* $\overline{z}$ *of* z. *Often we simply use the term "conjugate".*

Remark 7.5 *We mention in passing that other notations are used for the complex conjugate, including* z^*.

Exercise 7.15 *Show that a complex number* z *is real if and only if* $z = \overline{z}$. *Again this is much easier than it may seem at first glance!*

Definition 7.4 *If* $\mathbf{A}$ *is a matrix (which includes row and column vectors in* $\mathbb{C}^n$ *), the conjugate* $\overline{\mathbf{A}}$ *of* $\mathbf{A}$ *is defined by* $\left(\overline{\mathbf{A}}\right)_{ij} = \overline{a_{ij}}$. *That is, the conjugate of a matrix is obtained by conjugating each of its entries.*

Exercise 7.16 *Show that for any matrices* $\mathbf{A}$ *and* $\mathbf{B}$, $\overline{\mathbf{A}} + \overline{\mathbf{B}} = \overline{\mathbf{A} + \mathbf{B}}$ *and* $\overline{\mathbf{AB}} = \overline{\mathbf{A}}\,\overline{\mathbf{B}}$ *(assuming the operations are defined). Hint: work with the components.*

By Exercise 7.14, we have $z\overline{z} = a^2 + b^2$, which as we have observed is the square of the Euclidean distance of z to the origin. Therefore it makes sense geometrically to define

$$|z| = \sqrt{z\overline{z}}$$

The dot product in $\mathbb{R}^n$ was defined by $(v_1, \ldots, v_n) \cdot (w_1, \ldots, w_n) = \sum\limits_{\iota=1}^{n} v_i w_i$, and then we defined $\|\mathbf{v}\|^2 = \mathbf{v} \cdot \mathbf{v}$, which is a non-negative real number that extends the idea of "magnitude" from the one-dimensional magnitude $|x|$. The geometric interpretation of $\|\mathbf{v}\|$ is the "length" of $\mathbf{v}$. If we try the same definition with $\mathbb{C}^n$, the summands on the right are complex numbers, so the corresponding notion of "length" would be a complex number, which is geometrically problematic. "Length" should be a non-negative real number. Instead, we used what we learned about $\mathbb{C}$ and we define the *complex dot product*:

$$(z_1, \ldots, z_n) \cdot (w_1, \ldots, w_n) = \sum_{i=1}^{n} z_i \overline{w_i}$$

In the case of $\mathbf{z} \cdot \mathbf{z}$, the summands on the right are all $z_i \overline{z_i} = |z_i|$, which is a non-negative real number. In particular we have a non-negative real number $\|\mathbf{z}\| = \sqrt{\mathbf{z} \cdot \mathbf{z}}$, as hoped.

Notation 3 *There is some potential for confusion here, until you get used to the notation. First of all, remember that every real vector* $\mathbf{v} = (v_1, \ldots, v_n)$, *meaning that v_i is real for all i, is also a complex vector, since the real numbers are contained in the complex numbers. Since every real number is equal to its own conjugate, for real vectors* $\mathbf{v}$ *and* $\mathbf{w}$, *the real dot product is the same as the complex dot product (you will check this explicitly below). Therefore, from now on, to avoid confusion, we will use a different notation for the complex dot product, and will call it the complex inner product. There are way too many different notations for the complex inner product out there, including* $\langle \mathbf{v}, \mathbf{w} \rangle$, *which is what we will use, and notations that others use, such as* $\langle \mathbf{v} \mid \mathbf{w} \rangle$, $(\mathbf{v} \mid \mathbf{w})$, $(\mathbf{v}, \mathbf{w})$. *The other reason to use different notation is that the complex inner product has different properties from the real dot product, which as you will recall is symmetric, positive definite, and bilinear. I would encourage you to review those definitions before continuing.*

The complex inner product is a special case of a *Hermetian inner product* on a complex vector space. As such it satisfies similar properties to a symmetric bilinear form, with a twist. For example, the "sum" part of bilinearity still holds, but conjugation appears in the "scalar part". Conjugation also appears in the symmetry condition:

Definition 7.5 *A Hermetian inner product on a complex vector space V assigns to vectors* $\mathbf{v}$ *and* $\mathbf{w}$ *a complex number* $\langle \mathbf{v}, \mathbf{w} \rangle$ *with the following properties for all complex vectors* $\mathbf{u}, \mathbf{v}, \mathbf{w}$:

    1. $\langle \mathbf{v} + \mathbf{w}, \mathbf{u} \rangle = \langle \mathbf{v}, \mathbf{u} \rangle + \langle \mathbf{w}, \mathbf{u} \rangle$ *and* $\langle \mathbf{v}, \mathbf{w} + \mathbf{u} \rangle = \langle \mathbf{v}, \mathbf{w} \rangle + \langle \mathbf{v}, \mathbf{u} \rangle$

    2. $\langle k\mathbf{v}, \mathbf{u} \rangle = k \langle \mathbf{v}, \mathbf{u} \rangle$ *and* $\langle \mathbf{v}, k\mathbf{u} \rangle = \overline{k} \langle \mathbf{v}, \mathbf{u} \rangle$

3. *(Conjugate Symmetry)* $\langle \mathbf{v}, \mathbf{u} \rangle = \overline{\langle \mathbf{u}, \mathbf{v} \rangle}$

4. *(Positive definiteness)* $\langle \mathbf{v}, \mathbf{v} \rangle \geq 0$ *and* $\langle \mathbf{v}, \mathbf{v} \rangle = 0$ *if and only if* $\mathbf{v} = \mathbf{0}$.

Not everyone includes positive definiteness in the definition of Hermetian product, but instead speaks of a *positive definite* Hermetian product when the inner product satisfies this property.

Exercise 7.17 *Show that the complex inner product is a Hermetian inner product. You may use exactly the proof of the corresponding properties of the real dot product in this book, with appropriate modifications involving conjugates. So this is a great exercise to further your understanding of both topics.*

Exercise 7.18 *Show that if all of the entries of* $\mathbf{v} = (v_1, \ldots, v_n)$ *and* $\mathbf{w} = (w_1, \ldots, w_n)$ *are real, then* $\langle \mathbf{v}, \mathbf{w} \rangle = \mathbf{v} \cdot \mathbf{w}$.

Exercise 7.19 *Show this variation of positive definiteness that we will need below: If* $\mathbf{v} = (v_1, \ldots, v_n)$ *then* $\langle \mathbf{v}, \overline{\mathbf{v}} \rangle = \sum_{i=1}^{n} v_i^2$, *which is positive if* $\mathbf{v} \neq \mathbf{0}$.

Exercise 7.18 shows that the complex dot product is truly an extension of the dot product from $\mathbb{R}^n$ to $\mathbb{C}^n$. In fact, $\mathbb{R}^n$ is precisely the subspace of $\mathbb{C}^n$ of vectors with real entries (closure is clear, because the sum and product of real numbers is a real number).

Exercise 7.20 *Show by direct calculation that* $\langle \mathbf{v}, \mathbf{w} \rangle = \mathbf{v}^T \overline{\mathbf{w}}$.

We need to modify one more concept to the complex setting, namely the transpose:

Definition 7.6 *Let* $\mathbf{A}$ *be a complex matrix. The conjugate transpose, also known as the Hermetian transpose is the matrix* $\mathbf{A}^H = \overline{\mathbf{A}^T}$.

The basic properties only need small modification due to the involvement of the conjugate in scalar multiplication:

1. $(\mathbf{A}^H)^H = \mathbf{A}$

2. $(k\mathbf{A})^H = \overline{k}\mathbf{A}^H$.

3. $(\mathbf{A} + \mathbf{B})^H = \mathbf{A}^H + \mathbf{B}^H$

4. $(\mathbf{AB})^H = \mathbf{B}^H \mathbf{A}^H$

These should be simple for you to check now, and I won't even assign them! Also note that if $\mathbf{A}$ is a real matrix then $\mathbf{A}^H = \mathbf{A}^T$, so this is again a true extension of the idea of transpose to complex matrices. Applying the Fundamental Theorem of Algebra to the characteristic polynomial, we have the following theorem, which we will use without reference below.

Theorem 7.8 *Every complex square matrix* $\mathbf{A}$ *has at least one eigenvalue.*

Keep in mind that when $\mathbf{A}$ is real, $\mathbf{A}$ may have no *real* eigenvalues! But our first application is to show that the eigenvalues of a real *symmetric* matrix $\mathbf{A}$ are real. I won't state this as a separate theorem because it is a part of Theorem 7.10 in the next section. Suppose that λ is a (possibly complex) eigenvalue of a symmetric real matrix $\mathbf{A}$ with (possibly complex) eigenvector $\mathbf{v}$. Using the fact that, as a real matrix, $\mathbf{A} = \overline{\mathbf{A}}$, we may use Exercise 7.16 to calculate:

$$\mathbf{A}\overline{\mathbf{v}} = \overline{\mathbf{A}}\overline{\mathbf{v}} = \overline{\mathbf{A}\mathbf{v}} = \overline{\lambda\mathbf{v}} = \overline{\lambda}\overline{\mathbf{v}} \tag{7.8}$$

Next, using Exercise 7.20,

$$\overline{\mathbf{v}}^T \mathbf{A}\mathbf{v} = \overline{\mathbf{v}}^T (\lambda\mathbf{v}) = \overline{\mathbf{v}}^T (\lambda\mathbf{v}) = \lambda \langle \overline{\mathbf{v}}, \mathbf{v} \rangle$$

Recalculating the same quantity in a different way, using Equation (7.8) and symmetry of $\mathbf{A}$,

$$\overline{\mathbf{v}}^T \mathbf{A}\mathbf{v} = (\mathbf{A}^T \overline{\mathbf{v}})^T \mathbf{v} = (\mathbf{A}\overline{\mathbf{v}})^T \mathbf{v} = (\overline{\lambda}\overline{\mathbf{v}})^T \mathbf{v} = \langle \overline{\lambda}\overline{\mathbf{v}}, \mathbf{v} \rangle = \overline{\lambda} \langle \overline{\mathbf{v}}, \mathbf{v} \rangle$$

Eigenvectors are non-zero, and by Exercise 7.19, $\langle \overline{\mathbf{v}}, \mathbf{v} \rangle \neq 0$, and we conclude that $\lambda = \overline{\lambda}$. That is, λ is real.

So λ is real, what about the eigenvector $\mathbf{v}$ for λ? At this point you can just forget about the fact we had to use complex numbers to show that λ is real! We now have a real root to the characteristic polynomial of a real matrix $\mathbf{A}$, so you would find $\mathbf{v}$ in the same way that you always have—solving a system of linear equations with real coefficients, resulting in a real solution, i.e. a real eigenvector.

Here is another perhaps surprising application. You have already seen that the trace and determinant appear in the characteristic polynomial. We are now in a position to make a very precise connection between these quantities and eigenvalues. If $\mathbf{A}$ is a $n \times n$ matrix, the characteristic polynomial

$$p(t) = \det(t\mathbf{I} - \mathbf{A}) = t^n - tr\mathbf{A}t^{n-1} + \cdots + c_1 t + (-1)^n \det\mathbf{A} \tag{7.9}$$

completely factors into linear factors by the Fundamental Theorem of Algebra. That is,

$$p(t) = (t - \lambda_1)(t - \lambda_2) \cdots (t - \lambda_n)$$

where the constants λ_i are the roots, hence the eigenvalues of $\mathbf{A}$. (Don't forget that there might be some "repeats" in this list, since the eigenvalues might not all be different). Multiplying this back together, we get

$$p(t) = t^n - (\lambda_1 + \cdots + \lambda_n)t^{n-1} + \cdots + (-1)^n (\lambda_1 \lambda_2 \cdots \lambda_n) \tag{7.10}$$

Comparing the constant terms of (7.9) and (7.10) we reach the perhaps startling conclusion:

Theorem 7.9 *If* $\mathbf{A}$ *is an* $n \times n$ *matrix then its determinant is the product of its eigenvalues, and its trace is the sum of its eigenvalues.*

Of course some real matrices do not have real eigenvalues, but they always have complex eigenvaluses. One consequence of the above theorem is that the sum and product of the eigenvalues of a real matrix are always real.

7.6 The Spectral Theorem and SVD

Theorem 7.10 *If* $\mathbf{A}$ *is a symmetric matrix then all of its eigenvalues are real, and the eigenvectors corresponding to different eigenvalues are orthogonal.*

Proof. We proved that the eigenvalues are real in the previous section. Now suppose $\lambda_1 \neq \lambda_2$ are eigenvalues of $\mathbf{A}$, with eigenvectors $\mathbf{v}_1$ and $\mathbf{v}_2$, respectively. By definition, $\mathbf{A} = \mathbf{A}^T$, and by (4.17) and bilinearity we have

$$\lambda_1 (\mathbf{v}_1 \cdot \mathbf{v}_2) = \lambda_1 \mathbf{v}_1 \cdot \mathbf{v}_2 = \mathbf{A}\mathbf{v}_1 \cdot \mathbf{v}_1 = \mathbf{v}_1 \cdot \mathbf{A}^T \mathbf{v}_2 = \mathbf{v}_1 \cdot \mathbf{A}\mathbf{v}_2 = \mathbf{v}_2 \cdot \lambda_2 \mathbf{v}_2$$
$$= \lambda_2 (\mathbf{v}_1 \cdot \mathbf{v}_2)$$

Since $\lambda_1 \neq \lambda_2$, the only possible way that this equation can be true is for $\mathbf{v}_1 \cdot \mathbf{v}_2 = 0$. ∎

The third condition in Theorem 7.11 below may be restated as: $\mathbf{A}$ is *orthogonally diagonalizable*. Put another way, $\mathbf{A}$ may be diagonalized via an orthogonal matrix. This is a much stronger and more useful property than simply being diagonalizable.

Exercise 7.21 *Let* $\mathbf{A}$ *be the symmetric matrix* $\begin{bmatrix} 1 & 1 \\ 1 & 1 \end{bmatrix}$ *and* $\mathbf{P} = \begin{bmatrix} 0 & 1 \\ 1 & -1 \end{bmatrix}$. *Show by direct calculation that* $\mathbf{P}^{-1}\mathbf{A}\mathbf{P}$ *is not symmetric.*

Exercise 7.22 *Show that if* $\mathbf{A}$ *is symmetric and* $\mathbf{P}$ *is orthogonal then* $\mathbf{P}^{-1}\mathbf{A}\mathbf{P}$ *is symmetric.*

Theorem 7.11 *(Spectral Theorem) For an* $n \times n$ *matrix* $\mathbf{A}$, *the following are equivalent.*

1. $\mathbf{A}$ *is symmetric.*

2. $\mathbf{A} = \mathbf{P}^{-1}\mathbf{D}\mathbf{P}$, *where* $\mathbf{D}$ *is diagonal and* $\mathbf{P}$ *is orthogonal.*

3. $\mathbb{R}^n$ *has an orthonormal basis consisting of eigenvectors of* $\mathbf{A}$.

Moreover, the diagonal elements of $\mathbf{D}$ *are the eigenvalues of* $\mathbf{A}$, *with multiplicity. That is, if the dimension of the eigenspace of eigenvalue* λ *is* k, *then* λ *appears* k *times on the diagonal of* $\mathbf{D}$.

Proof. We will prove that the first statement implies the second statement by induction in n. The base case, $n = 1$, is easy–$[a]$ is already diagonal! (Strictly speaking, $\mathbf{P}$ is the orthogonal matrix $[1]$.) For the inductive step, we suppose that the statement is true for any $n \times n$ symmetric matrix and let $\mathbf{A}$ be an $(n + 1) \times (n + 1)$ symmetric matrix. Let λ be an eigenvalue of $\mathbf{A}$ with eigenvector $\mathbf{v}$ of length 1 (since eigenvectors are non-zero one can always divide by length to accomplish this). Using Gram-Schmidt, extend $\mathbf{v}$ to an orthonormal basis $S = \{\mathbf{v}, \mathbf{v}_2, \ldots \mathbf{v}_n\}$ of $\mathbb{R}^n$. The transition matrix from the standard basis to this new basis is $\mathbf{C}_S$, which has orthonormal columns and is therefore orthogonal. That is, $\mathbf{A} = \mathbf{C}_S^{-1} \mathbf{B} \mathbf{C}_S$, where $\mathbf{B}$ has block form

$$\begin{bmatrix} \lambda & \mathbf{0} \\ \mathbf{0} & \mathbf{C} \end{bmatrix}$$

By Exercise 7.22, $\mathbf{B}$ is symmetric, and therefore $\mathbf{C}$ is also symmetric. Since $\mathbf{C}$ is an $n \times n$ matrix, we may apply the inductive hypothesis: there is an orthogonal matrix $\mathbf{R}$ such that $\mathbf{C} = \mathbf{R} \mathbf{D}' \mathbf{R}^{-1}$ where $\mathbf{D}'$ is diagonal. It is an exercise to show that the matrix

$$\mathbf{S} = \begin{bmatrix} 1 & \mathbf{0} \\ \mathbf{0} & \mathbf{R} \end{bmatrix}$$

is orthogonal. The matrix

$$\mathbf{D} = \begin{bmatrix} \lambda & \mathbf{0} \\ \mathbf{0} & \mathbf{D}' \end{bmatrix}$$

is diagonal. By block multiplication, (Exercise 2.8),

$$\mathbf{S}^{-1} \mathbf{D} \mathbf{S} = \begin{bmatrix} 1 & \mathbf{0} \\ \mathbf{0} & \mathbf{R} \end{bmatrix}^{-1} \begin{bmatrix} \lambda & \mathbf{0} \\ \mathbf{0} & \mathbf{D}' \end{bmatrix} \begin{bmatrix} 1 & \mathbf{0} \\ \mathbf{0} & \mathbf{R} \end{bmatrix}$$

$$= \begin{bmatrix} 1 & \mathbf{0} \\ \mathbf{0} & \mathbf{R}^{-1} \end{bmatrix} \begin{bmatrix} \lambda & \mathbf{0} \\ \mathbf{0} & \mathbf{D}' \end{bmatrix} \begin{bmatrix} 1 & \mathbf{0} \\ \mathbf{0} & \mathbf{R} \end{bmatrix}$$

$$= \begin{bmatrix} \lambda & \mathbf{0} \\ \mathbf{0} & \mathbf{R} \mathbf{D}' \mathbf{R}^{-1} \end{bmatrix} = \begin{bmatrix} \lambda & \mathbf{0} \\ \mathbf{0} & \mathbf{C} \end{bmatrix} = \mathbf{B}$$

In other words,
$$\mathbf{A} = \mathbf{C}_S^{-1} \mathbf{B} \mathbf{C}_S = \mathbf{C}_S^{-1} \mathbf{S}^{-1} \mathbf{D} \mathbf{S} \mathbf{C}_S$$

Since the product of orthogonal matrices is orthogonal, setting $\mathbf{P} := \mathbf{S} \mathbf{C}_S$ finishes the proof of the second statement of the theorem.

Suppose the second statement is true and $\mathbf{D}$ is the diagonal matrix

$$\mathbf{D} = \begin{bmatrix} \lambda_1 & & & \\ & \lambda_2 & & \\ & & \ddots & \\ & & & \lambda_n \end{bmatrix}$$

that represents $T_{\mathbf{A}}$ with respect to some basis $\{\mathbf{v}_1, \ldots, \mathbf{v}_n\}$. Then by definition, $T_{\mathbf{A}}(\mathbf{v}_i) = \mathbf{D}\mathbf{v}_i = \lambda_i \mathbf{v}_i$, and therefore each $\mathbf{v}_i$ is an eigenvector with eigenvalue λ_i. To find an *orthonormal basis* of eigenvectors, simply use Gram-Schmidt to find an orthonormal basis for each eigenspace S_j corresponding to *distinct* eigenvalues $\lambda_{i_1}, \ldots, \lambda_{i_m}$. By definition, for any j, S_j is the subspace of all vectors having the same eigenvalue λ_{i_j}. This also shows the "moreover" statement of the theorem. Combining Theorem 7.10 and Theorem 6.22, Part 2, we see that the collection of all the individual orthonormal bases of the spaces S_j is an orthonormal basis for $\mathbb{R}^n$.

Rather than show directly that the third statement implies the first, we'll show the third implies the second and the second implies the first. Suppose the third statement is true and B is an orthonormal basis of eigenvectors. Since for each eigenvector $\mathbf{v}_i$ in B, $\mathbf{A}\mathbf{v}_i = \lambda_i \mathbf{v}_i$, this means that $\mathbf{A}$ is similar to a diagonal matrix with the eigenvalues along the diagonal, via the orthogonal transition matrix $\mathbf{C}_B$.

Finally, if $\mathbf{A}$ is orthogonally similar to a diagonal matrix, which is a symmetric matrix, then by Exercise 7.22, $\mathbf{A}$ is symmetric. ∎

Exercise 7.23 *Show that the matrix* $\mathbf{S} = \begin{bmatrix} 1 & 0 \\ 0 & \mathbf{R} \end{bmatrix}$ *in the proof of Theorem 7.11 is orthogonal.*

Theorem 7.12 *For any* $m \times n$ *matrix* $\mathbf{A}$*, the matrices* $\mathbf{A}^T\mathbf{A}$ *and* $\mathbf{A}$ *have the same domain for their corresponding linear transformations, as well as the same null space, row space, column space and rank. Moreover, if* $\mathbf{v}_1, \ldots, \mathbf{v}_n$ *is an orthonormal basis for* $\mathbb{R}^n$ *of eigenvectors of* $\mathbf{A}^T\mathbf{A}$*, then the eigenvalue for* $\mathbf{v}_i$ *is* $\|\mathbf{A}\mathbf{v}_i\|^2$*, which is non-negative.*

Proof. If $\mathbf{A}$ is an $m \times n$ matrix then $\mathbf{A}^T\mathbf{A}$ is an $n \times n$ matrix and therefore the domains $T_{\mathbf{A}}$ and $T_{\mathbf{A}^T\mathbf{A}}$ are both $\mathbb{R}^n$. If $\mathbf{X}$ is in the null space of $\mathbf{A}$, by definition this means that $\mathbf{A}\mathbf{X} = \mathbf{0}$. Then

$$\left(\mathbf{A}^T\mathbf{A}\right)\mathbf{X} = \mathbf{A}^T\left(\mathbf{A}\mathbf{X}\right) = \mathbf{A}^T(\mathbf{0}) = \mathbf{0}$$

That is, $\mathbf{X}$ is contained in the null space of $\mathbf{A}^T\mathbf{A}$.

Conversely, if $\mathbf{X}$ is a solution to $\left(\mathbf{A}^T\mathbf{A}\right)\mathbf{X} = \mathbf{0}$ then by Theorem 4.8, $\mathbf{X}$ is orthogonal to every row of $\mathbf{A}^T\mathbf{A}$. Since $\mathbf{A}^T\mathbf{A}$ is symmetric, $\mathbf{X}$ is also orthogonal to every vector in the column space of $\mathbf{A}^T\mathbf{A}$. By Theorem 6.12, the column space is the range of the linear transformation $T_{\mathbf{A}^T\mathbf{A}}$, which includes the vector $\mathbf{A}^T\mathbf{A}\mathbf{X}$. This tells us that $\mathbf{X} \cdot \mathbf{A}^T\mathbf{A}\mathbf{X} = 0$. Using (4.17) we have

$$0 = \mathbf{X} \cdot \left(\mathbf{A}^T\mathbf{A}\mathbf{X}\right) = \left(\mathbf{A}^T\mathbf{A}\right)^T \mathbf{X} \cdot \mathbf{X}$$

$$= \mathbf{A}^T(\mathbf{A}\mathbf{X}) \cdot \mathbf{X} = \mathbf{A}\mathbf{X} \cdot \mathbf{A}\mathbf{X} = (\|\mathbf{A}\mathbf{X}\|^2) \tag{7.11}$$

By positive definiteness, $\mathbf{A}\mathbf{X} = \mathbf{0}$. This completes the proof that $\mathbf{A}$ and $\mathbf{A}^T\mathbf{A}$ have the same null space.

Since $\mathbf{A}$ and $\mathbf{A}^T\mathbf{A}$ have the same domain and null space, and the row space is the orthogonal complement of the null space by Theorem 6.25, the row spaces of $\mathbf{A}$ and $\mathbf{A}^T\mathbf{A}$ must be equal. We may apply this result to the matrix $\mathbf{A}^T$ and see that $\mathbf{A}^T$ and $(\mathbf{A}^T)^T\mathbf{A}^T = \mathbf{A}\mathbf{A}^T$ have the same row space. But then $\mathbf{A} = (\mathbf{A}^T)^T$ and $\mathbf{A}^T\mathbf{A} = (\mathbf{A}\mathbf{A}^T)^T$ must have the same column space. Finally, since $\mathbf{A}$ and $\mathbf{A}^T$ have the same row space, they must have the same rank, which is the dimension of the row space.

For the last statement, using (4.17) we have

$$\|\mathbf{A}\mathbf{v}_i\|^2 = \mathbf{A}\mathbf{v}_i \cdot \mathbf{A}\mathbf{v}_i = \mathbf{v}_i \cdot \mathbf{A}^T\mathbf{A}\mathbf{v}_i = \mathbf{v}_i \cdot \lambda_i\mathbf{v}_i = \lambda_i\,(\mathbf{v}_i{\cdot}\mathbf{v}_i) = \lambda_i$$

∎

Exercise 7.24 *For each "=" in the calculation (7.11), state the theorems used.*

The final theorem in this text is of major importance in modern science and engineering, and involves all of the major aspects of linear algebra covered in this text. It may be viewed as an extention of the Spectral Theorem to non-square matrices.

Theorem 7.13 *(Singular Value Decomposition–SVD) Let $\mathbf{A}$ be an $m \times n$ matrix with rank k. Then*

$$\mathbf{A} = \mathbf{U}\mathbf{\Sigma}\mathbf{V}^T$$

where $\mathbf{V}$ and $\mathbf{U}$ are orthogonal and $\mathbf{\Sigma}$ is an $m \times n$ matrix with block form

$$\mathbf{\Sigma} = \begin{bmatrix} \mathbf{D} & \mathbf{0} \\ \mathbf{0} & \mathbf{0} \end{bmatrix}$$

where $\mathbf{D}$ is a $k \times k$ diagonal matrix whose diagonal elements $\sigma_1, \ldots, \sigma_k$, called the singular values of $\mathbf{A}$, are positive and in non-increasing order from top to bottom. The singular values are the square roots of the positive eigenvalues of $\mathbf{A}^T\mathbf{A}$.

The SVD of $\mathbf{A}$ is not unique, but the singular values are uniquely determined by $\mathbf{A}$. The existence of SVD (for certain types of matrices) was first discovered by mathematicians E. Beltrami and C. Jordan working in the area of differential geometry in the late 1800s. It seems they did not have practical applications in mind for their work. Certainly they could not possibly imagined that more than a century later their work would be used with powerful computers to solve computer engineering problems like image processing! The point is that beautiful mathematics (and this theorem is a beauty!), often turns out to be useful mathmatics, because what mathematicans regard as "beautiful" often reflects deep and previously hidden fundamental properties of our universe.

The geometric interpretation of SVD is that any linear transformation $T_\mathbf{A}$ is the composition of an orthogonal linear map (i.e. a rigid motion), followed by a linear map that re-scales coordinates either by the positive numbers σ_i or by 0, followed by another orthogonal linear map. The proof below provides one way to find an SVD in which $\mathbf{V}$ and $\mathbf{U}$ are fairly explicitly described. But the proof does not yield the most efficient algorithm to compute SVD, so I will not clutter the statement of the theorem with more details about the computational aspects.

Proof of SVD Theorem. For this final proof we will use several theorems, including the Spectral Theorem, without explicit reference. At this point, you should be able to fill in the references! The pace of this proof will be closer to that which you will see in a more advanced linear algebra course.

Since $\mathbf{A}^T\mathbf{A}$ is a symmetric matrix, $\mathbf{A}^T\mathbf{A}$ is orthogonally diagonalizable– that is there is a diagonal matrix $\mathbf{D}'$ with $\mathbf{D}' = \mathbf{V}\mathbf{A}^T\mathbf{A}\mathbf{V}^{-1} = \mathbf{V}\mathbf{A}^T\mathbf{A}\mathbf{V}^T$ for some orthogonal matrix $\mathbf{V}$. The diagonal elements $\lambda_1, \ldots, \lambda_m$, which are the eigenvalues of $\mathbf{A}^T\mathbf{A}$, may not be non-increasing order, but this can be fixed by row and column swaps, which correspond to left and right multiplication by orthogonal elementary matrices. This doesn't change the fact that we are multiplying on the left and right by orthogonal matrices in the final form, so we'll assume that the eigenvalues are in the proper order. Since the rank of $\mathbf{A}^T\mathbf{A}$ is the same as the rank of $\mathbf{A}$, which is the same as the rank of $\mathbf{D}'$, the first k eigenvalues are positive, and the rest are 0.

The columns $\{\mathbf{V}^1, \ldots, \mathbf{V}^n\}$ of $\mathbf{V}$ are orthonormal. We claim that the vectors $\mathbf{A}\mathbf{V}^i$ and $\mathbf{A}\mathbf{V}^j$ are orthogonal when $i \neq j$. In fact,

$$\mathbf{A}\mathbf{V}^i \cdot \mathbf{A}\mathbf{V}^j = \mathbf{V}^i \cdot \mathbf{A}^T\mathbf{A}\mathbf{V}^j = \mathbf{V}^i \cdot \lambda_j \mathbf{V}^j = \lambda_j \left(\mathbf{V}^i \cdot \mathbf{V}^j \right) = 0$$

Similarly, $\left\| \mathbf{A}\mathbf{V}^i \right\| = \sqrt{\lambda_i}$, which by definition is σ_i. Since the rank of $\mathbf{A}$ is k, and the column space of $\mathbf{A}$ is the range of $T_\mathbf{A}$, the set $B = \{\mathbf{A}\mathbf{V}^1, \ldots, \mathbf{A}\mathbf{V}^k\}$ is an orthogonal basis for the column space of $\mathbf{A}$. Now define $\mathbf{u}_i := \frac{\mathbf{A}\mathbf{V}^i}{\left\| \mathbf{A}\mathbf{V}^i \right\|} = \frac{1}{\sigma_k} \mathbf{A}\mathbf{V}^i$ for $1 \leq i \leq k$ so that $\{\mathbf{u}_1, \ldots, \mathbf{u}_k\}$ is an orthonormal basis for the column space of $\mathbf{A}$. Using Gram-Schmidt, we can extend to an orthonormal basis $B = \{\mathbf{u}_1, \ldots, \mathbf{u}_k, \ldots, \mathbf{u}_m\}$ of the range $\mathbb{R}^m$ of $T_\mathbf{A}$. Let $\mathbf{U} := \mathbf{C}_B$, let $\mathbf{D} = \sqrt{\mathbf{D}'}$ be the diagonal matrix with entries σ_i, and define $\boldsymbol{\Sigma}$ as in the statement of the theorem, filling in 0s as needed so that $\boldsymbol{\Sigma}$ is an $m \times n$ matrix. Then since $\mathbf{D}$ is diagonal we can compute

$$\mathbf{U}\boldsymbol{\Sigma} = [\sigma_1\mathbf{u}_1 \mid \cdots \mid \sigma_k\mathbf{u}_k \mid \mathbf{0} \mid \cdots \mid \mathbf{0}]$$

$$[\mathbf{A}\mathbf{v}_1 \mid \cdots \mid \mathbf{A}\mathbf{v}_n] = \mathbf{A}\mathbf{V}$$

Since $\mathbf{V}$ is orthogonal, this implies $\mathbf{A} = \mathbf{U}\boldsymbol{\Sigma}\mathbf{V}^T$. ∎

MP 75 *SVD Decomposition*

Index

0-row, 3

$\mathbb{R}^\infty$, 115

$\mathbf{A}_i$, 8
$\mathbf{A}^j$, 8
$\mathbf{A}_B$, 147
$\mathbf{A}_{B_V B_W}$, 146
$\mathbf{A}^H$, 183
$\langle \mathbf{v}, \mathbf{w} \rangle$, 182
alternating sum, 71
angle between vectors, 82, 84
augmented matrix, 21
    reduced form of, 28
    system of linear equations, 40

backsolving, 49
base case, 70
Basis Extension Theorem, 136
bilinearity of dot product, 87
block matrix, 44
block matrix multiplication, 44

$\mathbb{C}$, 179
$\mathcal{C}$ (continuous real functions), 114
cancellation law, 20
Cauchy-Schwarz Inequality, 85
characteristic polynomial, 166
closure
    addition and scalar
        multiplication, 117
codomain, 101, 120
coefficient matrix, 28
coefficients, 24
column operations, 53
column space, 150
commutative law, 10
complex conjugate, 181

complex dot product, 182
complex inner product, 182
component free calculation, 86
components
    of a vector, 76
composition of functions, 107
conjugate transpose, 183
consistent system, 29
constant term, 24
contact matrix, 36
continuous functions
    vector space of, 114
coordinate free, 83
coordinates
    of a point, 76
coordinates of a vector
    with respect to a basis, 129, 145
Cramer's Rule, 74
cross product, 98
$\mathbf{C}_S$, 100

$\mathcal{D}$, 117
D1-D3, 59
D4, 60
degenerate parallelogram, 108
δ_{ij}, 156
determinant
    2×2 matrix, 65
    muiltilinearity of, 139
    properties of, 59, 60
diagonal matrix, 2
differentiable functions
    space of, 117
dilation, 102
dimension, 134
Dimension Sum Theorem, 154
direction

of a vector, 78
direction of a vector, 85
direction of motion, 78
disease spread, 36
domain, 101, 120
dot product
    general properties, 88
    of column and row vectors, 8
    of vectors, 80
dot product (scalar product), 83
Duck
    walks like and quacks like, 67

echelon form, 49
eigenspace, 168
eigenvalue, 165
eigenvector, 165
elementary matrix, 50
Euclidean space, 75

$\mathbb{F}_2$, 12
field, 11
free column, 30
free variable, 30
Fundamental Theorem of Algebra,
    180

Gauss-Jordan Elimination, 54
Gauss-Jordan elimination
    outcomes, 29
Gram-Schmidt process, 161, 162

Hadamard product, 7
Hermetian inner product, 182
Hermetian transpose, 183
homogeneous matrix equation, 41
homogeneous system, 41

identity matrix, 16
image of a function, 101
inconsistent system, 29
induction
    proof by, 186
infinite dimensional, 134
inhomogeneous matrix equation, 41
inhomogeneous system, 41

inner product, 160
integers, 13
intersection of sets, 119
intersection of subspaces, 119
inverse matrix, 16
    properties, 23
invertible matrix, 16
isometry, 159
iterative step, 70

kernel
    of a linear transformation, 101,
        120
Kronecker Delta, 156

Lagrange's identity
    cross product, 99
Law of Cosines, 82
leading entry, 3
leading variable, 30
length
    of a vector, 83
    of a vector in the plane, 81
line
    parametric equations for, 91, 93
    vector equation for, 90
linear combination, 121
linear dependence and independence,
    123
linear equation, 25
linear equations
    system of, 26
linear isomorphism, 132
linear system
    homogeneous, 41
    matrix for, 27
    solving with Gauss-Jordan
        elimination, 29
linear transformation, 102, 120
linear transformation of a matrix,
    101
lower triangular matrix, 2
LU-factorization, 56

$\mathbb{M}_{m,n}$, 115
magnitude

of a vector, 78
matrices
    vector space of, 115
matrix, 1
    augmented, 21, 27
    column and row notation, 8
    diagonal, 2
    diagonal of, 2
    elementary, 50
    entry notation, 2
    entry or component of, 2
    inverse of, 16, 17
    invertible, 16
    lower triangular, 2
    negation of, 13
    non-singular, 23
    of a system of linear equations,
        28
    singular, 23
    size, 1
    skew symmetric, 20
    square, 1
    stochastic, 36
    symmetric, 2, 20
    trace of, 17
    transpose of, 19
    upper triangular, 2
    zero, 6
matrix addition, 7
    negatives and **0**, 14
matrix multiplication
    Associative Law, 14
    compatible size, 7
    Distributive Law, 15
    entry expression, 9
    summation expression, 18
    zero matrix, 17
matrix representation
    with respect two bases, 146
multilinearity
    of determinant, 139
multiplicity
    geometric, of an eigenvalue, 177

necessity in a proof, 174

network analysis, 37
non-singular matrix, 23
norm, 83
normal vector
    to a line in the plane, 95
notation flexibility, 2, 101
null space, 153

one-to-one, 132
onto function, 132
orthogonal
    subspaces, 159
    vectors, 93
orthogonal matrix, 157
orthogonal projection, 97
    onto a subspace, 162
orthogonally diagonalizable, 185
overdetermined system, 46

$\mathcal{P}$, 116
$\mathcal{P}(n)$, 117
parallel argument, 81
parallel vectors, 94
Parallelogram Law, 81
parallelogram spanned by plane
        vectors, 79
parallepiped, 110
parameterized equations, 47
parameters, 47
parametric equations for a line, 91,
        93
parametrzed solution, 48
particular solution, 41, 47, 48
pivot, 49
plane
    in $\mathbb{R}^3$, 96
polarization formula, 88
polynomials, 116
    standard basis, 127
polynomials of degree $\leq n$
    vector space of, 117
population migration, 35
position function, 76
position vector, 90
positive definite property

of the norm, 83
$proj_\mathbf{v}(\mathbf{u})$, 97
proof by contradiction, 43
proof by induction, 169, 186
proper subset, 123

$\mathbb{R}^\infty$, 137
range
    of a function, 101
    of a linear transformation, 120
rank, 150
rational numbers, 12
reduced echelon form, 3
    3×3 matrix, 6
reduced form
    augmented matrix, 28
reduced form outcomes, 30
reparameterization function, 92
ρ_θ, 105
right hand rule
    cross product, 99
rigid motion, 159
$\mathbb{R}^n$, 75
RO1-RO3, 4
rotation clockwise by angle θ, 105
rotation matrix, 106
rotations in space, 10
row additivity, 138
row operations, 4
row space, 150

scalar, 1
scalar multiplication, 7
    of matrices, 17
    of vectors, 79, 83
scalar product (dot product), 80
scalar triple product, 99
similar matrices, 148
singular matrix, 23
size of a matrix, 1
skew symmetric matrix, 20
solution of a system of llinear
        equations, 26
span
    of a set of subspaces, 159

span of vectors, 121
Spanning Set Reduction Theorem,
    138
standard basis
    for $\mathbb{M}_{2\times 2}$, 129
    in $\mathbb{R}^3$, 98
    for $P(n)$, 127
    in $\mathbb{R}^n$, 100
stochastic matrix, 36
subscript notation with comma, 32
subspace, 116
sufficiency in a proof, 174
sum of vectors, 83
symmetric bilinear form, 160
symmetric matrix, 2, 20
symmetry of dot product, 88
system matrix, 28
system of linear equations, 26
    consistent, 29
    inconsistent, 29
    solutions of, 26

trace, 17
    summation property, 18
transistion matrix, 145
transpose, 19
triple vector product, 99
trivial solution, 41

underdetermined system, 46
underlying matrix, 44
union of subspaces, 119
uniqueness theorems, 64
unit vector, 78, 85
upper triangular matrix, 2

$\frac{\mathbf{v}}{\|v\|}$ notation, 85
vector
    length or norm, plane, 81
    representation of (plane), 75
vector addition, 78
vector space, 113
vectors
    addition of, 78
    column and row, 8
velocity vector, 76

volume, 111

well-defined function, 61
WOLOG, 95
Wronskian, 127

$\bar{z}$ (complex conjugate), 181
zero divisor, 20
zero matrix
    multiplication by, 17

For Product Safety Concerns and Information please contact our EU
representative GPSR@taylorandfrancis.com
Taylor & Francis Verlag GmbH, Kaufingerstraße 24, 80331 München, Germany